Enhanced Oil Recovery Processes - New Technologies

Edited by Ariffin Samsuri

Published in London, United Kingdom

IntechOpen

Supporting open minds since 2005

Enhanced Oil Recovery Processes - New Technologies
http://dx.doi.org/10.5772/intechopen.77411
Edited by Ariffin Samsuri

Contributors
Peyman Pourafshary, Nikoo Moradpour, Afeez Gbadamosi, Radzuan Junin, Muhammad Manan, Augustine Agi, Jeffrey Oseh, Rini Setiati, Septoratno Siregar, Deana Wahyuningrum, Nasser Al Hinai, Hisham Ben Mahmud, Shattia Arumugam, Walid Mahmud, Erfan Mohammadian, Amin Azdarpour, Hossien Hamidi, Nur Hidayati Binti Othman, Aqilah Dollah, Siti Nurliyana Binti Che Mohamed Hussein, Rozana Azrina Binti Sazali, Badrul Mohamed Jan, Baozhen Li, Jian Zhang, Fengjiu Zhang, Xiaodong Kang

First published in London, United Kingdom, 2019 by IntechOpen
IntechOpen is the global imprint of INTECHOPEN LIMITED, registered in England and Wales, registration number: 11086078, 7th floor, 10 Lower Thames Street, London, EC3R 6AF, United Kingdom
Printed in Croatia

British Library Cataloguing-in-Publication Data
A catalogue record for this book is available from the British Library

Additional hard and PDF copies can be obtained from orders@intechopen.com

Enhanced Oil Recovery Processes - New Technologies
Edited by Ariffin Samsuri
p. cm.
Print ISBN 978-1-78985-107-6
Online ISBN 978-1-78985-108-3
eBook (PDF) ISBN 978-1-83968-153-0

For EU product safety concerns:
IN TECH d.o.o., Prolaz Marije Krucifikse Kozulić 3, 51000 Rijeka, Croatia, info@intechopen.com or visit our website at intechopen.com.

Meet the editor

Professor Dr Ariffin Samsuri is a Petroleum Engineering Senior Professor at the University of Technology (UTM) and he has vast experience in managing R&D projects as well as the faculty and department. He has more than 39 years of teaching and supervising experience in petroleum engineering including academic program development, staffing, faculty, and facilities establishment. He has also been appointed as a visiting professor at the University of Technology Petronas, as a peer reviewer, external examiner, expert, and academic advisor. He has published more than 147 technical papers for conferences/seminars and journals, authored 6 books, 4 book chapters, translated 4 books, edited 12 books, 5 research monographs, and reviewed 150 papers. He has also been involved in more than 37 research projects in drilling and production optimization, biofuel, and nanotechnology application in oil and gas.

Contents

Preface

Concerned with the existing oilfield production decline, the shortage of new oil reserves, and increasing world energy demand, the oil sector continues to search for economic and efficient techniques to enhance their oil recovery from the oil field using several EOR methods. Despite its high efficiency, widely acclaimed potentials, and limitations, the LSWF, hybrid, and nanotechnology applications have gained vast interest with promising future to increase ultimate oil recovery, tackle operational challenges, reduce environmental damage, and allow the highest feasible recoveries with lower production costs. This synergistic combination has opened new routes for new materials with fascinating properties. This book intends to provide an overview of some EOR technology such as LSWF, hybrid, and nanotechnology application in EOR processes.

This book contains seven chapters and it aims to provide readers with a comprehensive overview of the latest advancements in EOR processes and technologies such as China Offshore Chemical EOR scenario, direct gas thickener, surfactant flooding, LSWF, hybrid EOR, CO_2-EOR, and nanotechnology application in EOR. The seven chapters have been selected because they represent areas of EOR technology for which sufficient technical progress has been achieved to result in technological advancement.

I have enjoyed working with the authors of this book, who have been most diligent in preparing their chapters. I wish to thank them for their contribution, patience, and commitment in the process. Each chapter is designed to help the reader gain insight on the most important aspects of each topic.

On behalf of the authors, I would like to extend our heartfelt gratitude to IntechOpen and Sara Bacvarova, Author Service Manager, for their support and to all the staff who have provided input, drafted, revised, and produced this book.

It is my hope that this book is used as a source of knowledge and technology, by all concerned about the better future of the oil sector and world energy supply demand.

Ariffin Samsuri
University of Technology Malaysia,
Malaysia

Chapter 1

Hybrid EOR Methods Utilizing Low-Salinity Water

Peyman Pourafshary and Nikoo Moradpour

Abstract

Low-salinity water (LSW) flooding has been applied in sandstone and carbonate formations to improve oil recovery. Wettability alteration by LSW has been identified as the dominant driving mechanism for the incremental oil recoveries. LSW flooding has been combined with other EOR methods to develop new hybrid approaches to improve crude/brine/rock (CBR) interactions with the objective of overcoming some of the LSW flooding downsides, which include oil trapping and fine migration. Hybrid methods can provide higher oil recovery than each stand-alone technique. For instance, changes in gas solubility during LSW injection positively affect the performance of LSW/gas hybrid injection. LSW/surfactant flooding can contribute to incremental recovery by simultaneously lowering interfacial tension (IFT) and wettability alteration. The synergistic effect of fluid redistribution by LSW and enhanced water mobility by polymer flooding improves oil detachment and displacement in porous media through the application of the hybrid approach LSW/polymer flooding. Nanoparticles (NPs), mainly SiO_2, can alter wettability toward more water wetness in combination with LSW, and hot LSW can improve heavy oil production by reducing viscosity. Hence, the synergistic effect of hybrid EOR methods based on LSW flooding is considered a novel EOR approach to improve oil recovery.

Keywords: hybrid EOR, LSW, WAG, surfactant, polymer, nanofluid, hot water, wettability, IFT

1. Introduction

Waterflooding is the most widely used method to increase oil recovery. Recent studies show that in many cases, by modifying the salinity and ionic content of the injected water, oil recovery by waterflooding improves. The injection of water with lower salinity or adjusted ion composition triggers different mechanisms that modify the wettability of sandstone and carbonate formations. The bonding of polar components in the crude oil with the carbonate or sandstone rock surface is affected by the salinity and composition of the injected water, which generally produces wettability alteration of the rock surface. Several mechanisms have been proposed in the literature for low-salinity water flooding EOR, such as fine migration, rock dissolution, pH increase, multicomponent ion exchange, and double-layer expansion. The combination of these mechanisms is believed to affect the oil recovery in carbonate and sandstone formations. For more information on LSW and the governing mechanisms, the reader is referred to [1].

The idea of combining two (or more) EOR methods, known as hybrid methods, has been investigated recently to promote the activation of several oil recovery mechanisms to increase the ultimate oil recovery, tackle operational challenges, reduce environmental damage, and lower the production costs. Hybrid methods can be optimized for different injection scenarios to achieve the highest feasible recoveries. LSW flooding has been found to be effective when combined with gas injection (mainly CO_2), surfactant and/or polymer flooding, nanofluid injection, and hot water injection, each of which can improve the oil recovery through several mechanisms such as mobility control, wettability alteration, IFT reduction, etc. Experimental and modeling studies reviewed in this chapter have found that LSW hybrid methods can provide up to 30% original oil-in-place (OOIP) incremental oil recovery.

2. LSW/gas hybrid EOR technique

The application of hybrid LSW/gas flooding has recently attracted the attention of different researchers. Various injection schemes have been studied using experimental and simulation approaches. LSW can be injected into water alternating gas (WAG) or simultaneous water alternating gas (SWAG) modes. Moreover, gas may be injected after completion of LSW injection to improve the total oil recovery. CO_2 is generally used as the injection gas in this hybrid approach as it is cheap, has a lower minimum miscibility pressure (MMP), and provides environmentally positive results. Injection of LSW affects the solubility of gas in water and, consequently, the gas/oil interaction and oil recovery. There are two different observations during the hybrid injection of gas and LSW. Some researchers reported benefits from the application of this hybrid method, while others did not observe any incremental recovery compared to continuous gas injection (CGI).

Experimental and modeling studies show that the solubility of gas, especially CO_2, in brine increases with decreasing salinity of water due to the salting out effect [2–5]. Hence, higher CO_2 solubility results in lower amount of free gas available to come into contact with the oil and improve oil mobility, which reduces the oil recovery. On the other hand, many other experiments resulted in higher oil recovery at lower water salinities, as higher solubility increases the gas diffusion in water, which affects the waterfront, mobility ratio, and stability of the injection fluid. Improvement in these parameters leads to lower fingering of the injected fluid that increases its contact with previously bypassed oil droplets in the porous media. Therefore, dissolved gas in water changes the shape of the waterfront which contacts unswept oil more easily than free gas.

In both carbonate and sandstone formations, hybrid LSW/gas injection can be beneficial if it can alter the mobility of the injected fluid. Aleidan and Mamora [6] investigated different schemes of water/CO_2 injection such as SWAG and WAG at different salinities to study the effect on oil recovery in carbonate core samples. They observed higher oil recovery for both injection schemes when switching to LSW, as shown in **Figure 1**. More dissolution of gas in brine reduces the mobility of the injected fluid. An important observation in this study was that the LSW alone was not effective; hence, the observed incremental oil recovery was due to the hybrid method and the synergy between gas injection and LSW injection.

The same trend was observed by Jiang et al. [7]. They investigated the effect of water salinity on the performance of WAG during miscible flooding in sandstones using a highly viscous crude oil. Higher solubility of gas in low-salinity water controls the mobility of the water and reduces the mobility ratio of water and viscous oil. Hence, they observed better oil recovery, as shown in **Figure 2**.

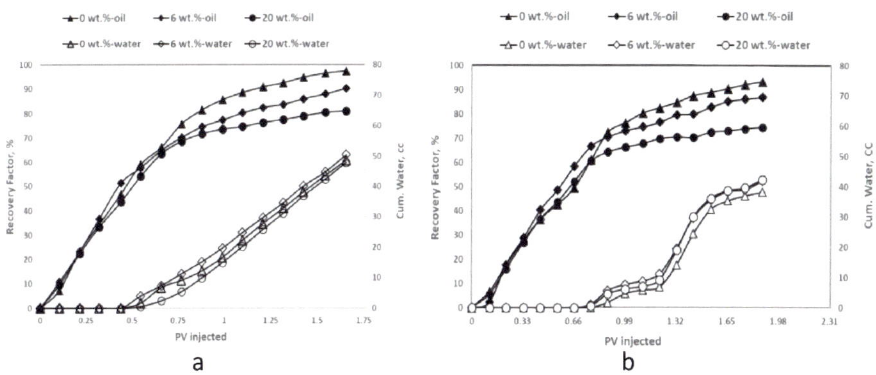

Figure 1.
Recovery factor of oil and cumulative water production at different salinity levels (0, 6, and 20%) (i.e., 0 wt.%-oil and 0 wt.%-water are oil production and water production by water injection with salinity of 0%, respectively): (a) SWAG, (b) WAG [6].

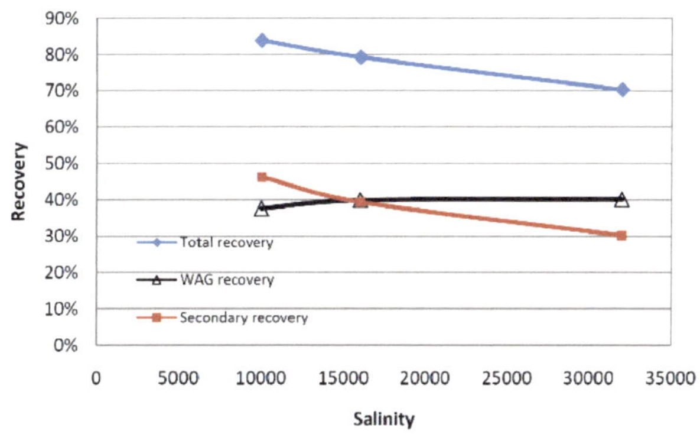

Figure 2.
The effect of brine salinity (in ppm) during hybrid LSW/gas WAG injection on oil recovery at secondary (waterflooding) and tertiary modes [7].

Kumar et al. [8] experimentally studied the effect of hybrid smart water/CO_2 flooding into sandstone samples with high clay content. 0.5 NaCl, 0.5 KCl, and 0.5 wt% $MgCl_2$ solution were used as smart brines for the hybrid method. They observed that smart water alternating gas injection controlled the mobility ratio of the fluids injected and reduced fluid channeling and oil bypassing effects. Hence, smart water alternating immiscible gas flooding recovered more oil than the stand-alone smart water and continuous gas injection (CGI) in their experiments. It should be noted that the rock samples in their study were initially water wet; therefore, wettability alteration by smart water was not the dominant mechanism during the hybrid EOR.

If the initial mobility ratio of water and oil is favorable, application of the hybrid method does not provide any benefit, as it does not affect the flow front stability. In these cases, salting out effect becomes more dominant, which reduces the amount of free gas contacting oil during the hybrid method. Hence, switching to LSW alternating gas (LSWAG) reduces the oil recovery. In [7], the LSWAG method was used to improve the recovery of a model oil composed of *n*-decane and *n*-hexadecane in sandstones. Due to the low viscosity of the oil and the homogeneity of the small artificial cores used in their experiments, the initial sweep efficiency was high, and LSWAG did not have a noticeable effect on oil recovery as shown in **Figure 3**. Hence, LSWAG is not recommended at these conditions.

Another factor that should be considered for an effective hybrid LSW/gas method is the performance of CGI. Wherever the recovery by stand-alone gas injection is high, application of the hybrid method does not provide noticeable incremental oil. For example, Al-Shalabi et al. modeled the hybrid injection of LSW and miscible CO_2 into carbonates using UTCOMP simulator. In their studies, the recovery factor by CGI was 98.9%, and shifting to the hybrid method just changed it to 99.7% by viscous fingering control [9]. Hence, in similar conditions, where the dominant mechanism is the miscibility of the gas, application of the hybrid method is not recommended.

An important parameter that affects the success of the hybrid LSW/gas method is the initial wettability of the rock. Wettability alteration from oil wet to water wet is considered as one of the main reasons for the positive performance of LSW, especially in sandstones [10]. Hence, if the initial wettability of the rock is water wet, LSW and consequently the hybrid method do not work. In this condition, the salting out effect controls the recovery mechanism during hybrid injection. Ramanathan et al. experimentally studied seawater alternating gas (SeaWAG) and LSWAG injection to recover oil from water-wet sandstone [11]. The recovery factor in LSWAG was lower than SeaWAG as the rocks were initially strongly water wet. Conversely, in an aged oil wet core, recovery by WAG changed from 76 to more than 97% when the water utilized changed from seawater to low-salinity brine, as shown in **Figure 4**.

The importance of the initial wettability was also confirmed by Teklu et al. [1]. They did several contact angle measurements for carbonate, sandstone, and shale samples at different initial wettability conditions, as shown in **Figure 5**, in which, case A shows the initial contact angle of an oil-wet disk, while case D shows the original water-wet condition of rock before aging by the oil. B and C cases show alteration in the contact angle by sample aging in seawater and CO_2 and in low-salinity brine and CO_2, respectively. LSW/CO_2 is more effective in the presence of CO_2 because the wettability of the rock changes toward a more hydrophilic surface. As shown, this mechanism is more effective when the initial wettability is oil wet.

In [13], Al-Abri et al. experimentally studied the hybrid immiscible CO_2 and smart water injection into sandstone core samples. They investigated the synergistic effects between gas injection and different ions in the water samples. Three synthetic brines were used in their work which contained 5000 ppm NaCl, $MgCl_2$, and KCl, respectively. Considerable improvement in oil recovery was observed, as shown in **Figure 6**. The maximum solubility of CO_2 in brine was observed in the

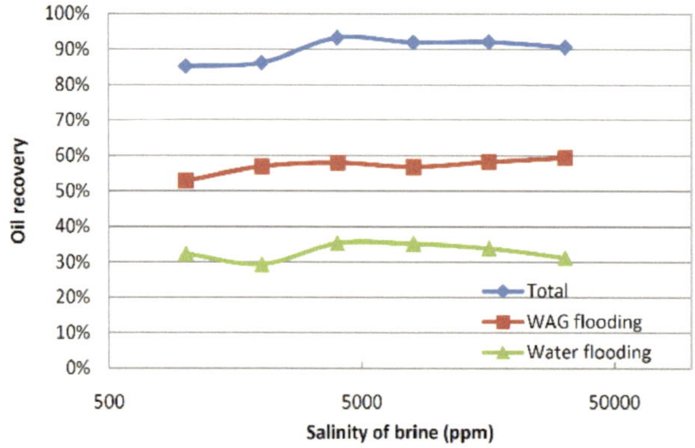

Figure 3.
The effect of salinity during hybrid LSWAG injection on recovery of low-viscosity oil (green curve is for secondary water flooding, red curve is for LSWAG, and blue curve is the total recovery) [7].

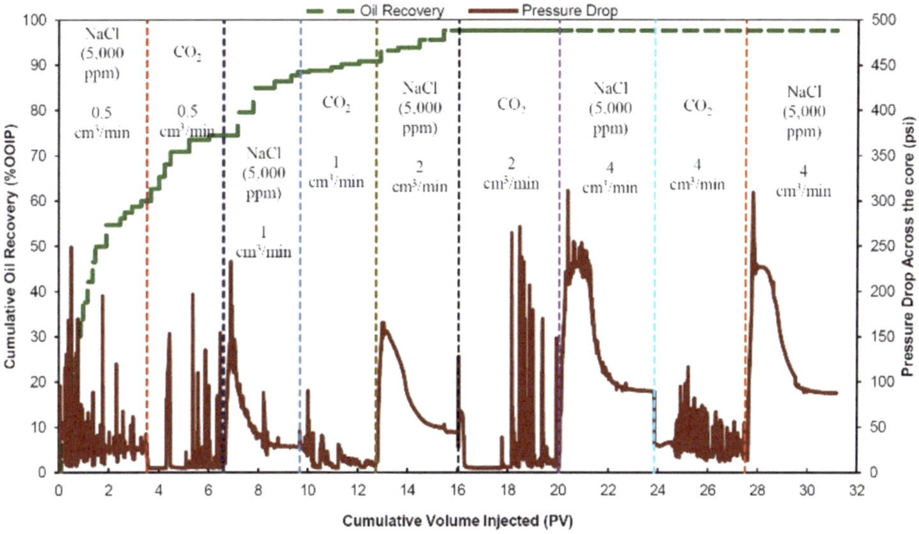

Figure 4.
Oil recovery and pressure drop across an oil-wet core during LSW alternating CO_2 [12].

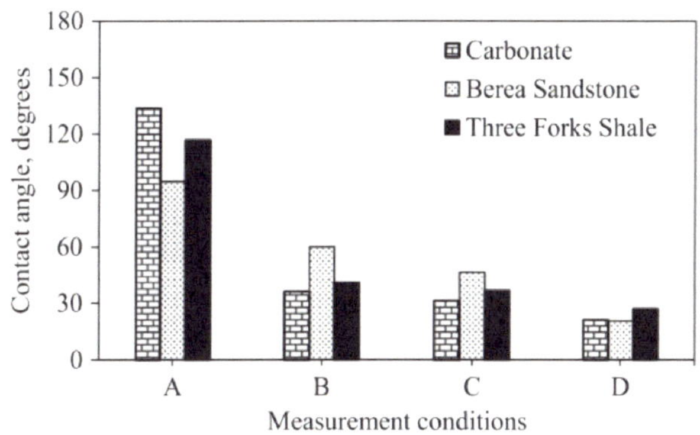

Figure 5.
Contact angle alteration by sample aging in high-salinity brine/CO_2 and low-salinity brine/CO_2 [1].

water containing $MgCl_2$, which also showed the lowest oil recovery among the three tests. In their work, smart water samples were effective, altering the wettability of the rock to more water wet without gas injection due to multicomponent ion exchange.

Hence, generally, wherever the LSW alone is effective, the hybrid method shows good performance and provides higher oil recovery than CGI and high-salinity water alternating gas. This point was also observed and confirmed by [14, 15]. AlQuraishi et al. [14] showed that low-salinity alternating miscible CO_2 method was useless for clay-free sandstones, but wherever clays were present, the recovery value was 35.1% of the OOIP for LSWAG.

A study by Yang et al. [16] showed that at a constant pressure and temperature conditions, the presence of CO_2 reduces the oil/brine IFT. Hence, this may also be considered as one of the mechanisms for incremental oil recovery by the hybrid method. References [1, 12] observed a reduction in IFT of less than 10 dynes/cm. On the other hand, in [8], higher IFT in the presence of CO_2 was reported. It should be noted that the alteration in IFT is not significant and cannot be considered as a dominant mechanism during the hybrid LSW/gas approach. In [17], Bennion

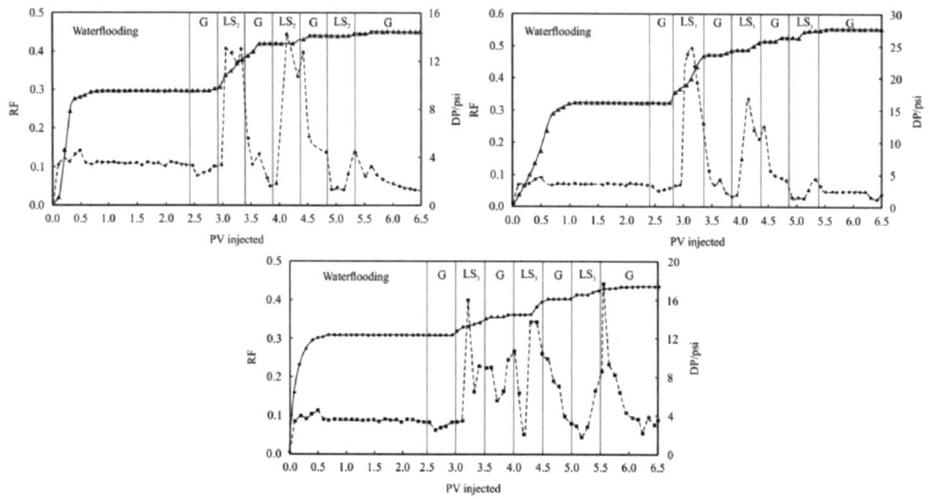

Figure 6.
Oil recovery and pressure drop for hybrid smart water alternating CO_2 flooding. (a) 5000 ppm $MgCl_2$ (top left), (b) 5000 ppm NaCl (top right), and (c) 5000 ppm KCl (bottom) [13].

et al. experimentally concluded that the increase in CO_2 solubility in low-salinity brine reduces the IFT of CO_2 and brine, which may lead to wettability alteration. Therefore, it can be considered that wettability alteration and mobility control are the dominant mechanisms for hybrid LSW/gas methods. In cases in which the initial wettability is water wet, the LSW stand-alone is not effective. In these situations, mobility control is the mechanism that shows the most important effect in improving oil recovery.

Different injection schemes have been applied to study the benefits of hybrid methods. Besides WAG and SWAG, the injection of CO_2 after LSW flooding provides extra oil recovery and can be considered as a novel approach for application in oil fields. Teklu et al. observed more than 20% incremental oil recovery by this approach [1].

Hybrid methods are also beneficial in terms of decreasing the operational costs of enhanced oil recovery processes. Previous research [9] has shown that the application of simultaneous LSW alternating gas leads to faster production of oil, as shown in **Figure 7**. This occurs due to the alteration of the reservoir rock wettability that increases the oil relative permeability. Also, the application of hybrid gas/LSW methods reduces the gas utilization factor. Kulkarni et al. observed a lower gas utilization factor during LSW alternating miscible CO_2 flooding [18]. However, this issue should be considered carefully during immiscible flooding, because the higher solubility of CO_2 in LSW requires more gas to make contact with the oil, which increases the gas utilization factor. Thus, more experimental and modeling studies are required in this area.

Most of the previous research in this area has focused on the performance of hybrid methods in sandstones. Analysis of the special interaction of LSW and gas with carbonate rock needs more investigation. For example, rock dissolution is considered as an effective mechanism during LSW injection in carbonates. Hence, geochemical analysis is essential to study the hybrid LSW/gas injection in rocks with high calcite and dolomite content. An earlier work [19] has performed simulations to compare the geochemical analysis results of LSW, CGI, and hybrid methods for different types of carbonates. Results indicated that hybrid methods can accelerate the dissolution process especially for high dolomite concentrations. Consequently, comprehensive experimental studies are required to investigate the

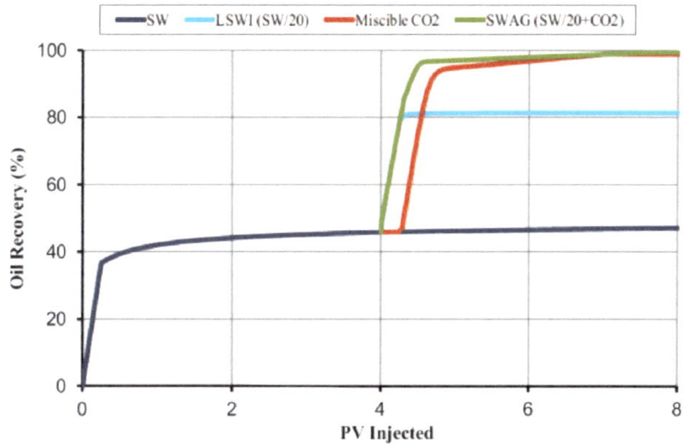

Figure 7.
Faster production of oil by LSW alternating gas injection than CGI [9].

geochemical parameters during hybrid LSW/gas injection methods to explain the oil recovery mechanisms of these processes.

Up to now, most of the work conducted in this area has been experimental research. There are few simulation studies in this field. Dang et al. [20] simulated a hybrid LSW alternating miscible CO_2 flooding injection in a 1D heterogeneous core and then upscaled the model to simulate the process at field scale. Their study showed that the hybrid approach overcomes the WAG late production problem. We recommend a comprehensive field-scale simulation study based on experimental work to analyze the practical benefit of hybrid approaches.

3. LSW/polymer hybrid EOR technique

Polymer flooding, as a well-known and effective EOR method, can also be considered as a complementary method to enhance the capability of LSW flooding as a hybrid approach. Polymer flooding affects the macroscopic sweep efficiency in porous media by improving the displacement mobility ratio. On the other hand, LSW flooding affects the microscopic sweep efficiency by changing CBR interactions and wettability. Hence, the hybrid low-salinity/polymer flooding provides the benefits of both methods and can be considered as a novel EOR approach.

Different injection schemes were experimentally studied and modeled in the literature. LSW can be injected as a preflush before polymer flooding, or polymer can be injected prior to LSW. The first approach is more effective than the second one because the injection of LSW changes the wettability to more water wet, which alters the distribution of the remaining oil saturation in the porous medium. Oil droplets are detached from small pores and accumulated in bigger and middle-sized pores. Hence, the injection of polymer recovers the redistributed oil more easily. Torrijos et al. [21] experimentally investigated the synergy between LSW flooding and polymer flooding in sandstones. Both injection modes were studied by injecting a low-salinity polymer (LSP), which was prepared by dissolving 1000 ppm hydrolyzed polyacrylamide (HPAM) in 1000 ppm NaCl brine as LSW. **Figure 8** shows that the hybrid LSW/LSP provides around 20% higher total oil recovery than the LSP/LSW method.

A similar trend was observed by Alsofi et al. [22] who studied the synergy between LSW/polymer flooding in a carbonate formation. In this work, heavy crude oil, high-salinity (69,000 mg/L) brine, and low-salinity brine (6900 mg/L)

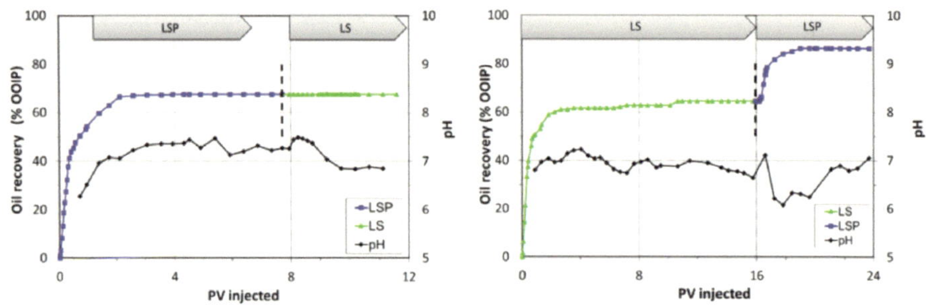

Figure 8.
Oil recovery by two combinations of polymer flooding and LSW flooding [21].

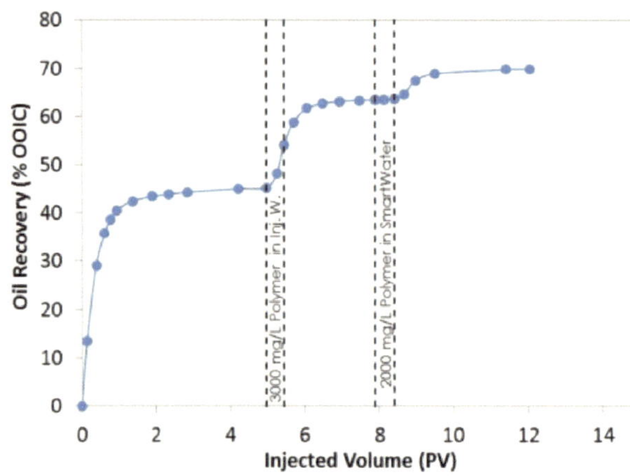

Figure 9.
Recovery profile for tertiary hybrid flooding method [22].

were used. After secondary waterflooding, polymer flooding followed by polymer-LSW flooding provided more than 24% OOIP additional recovery, as shown in **Figure 9**. The same findings were reported by [23–25].

Hence, the sequence of injection affects total oil recovery. Almansour et al. experimentally investigated the hybrid LSW/polymer flooding method in Berea and Bentheimer sandstone samples. Persian Gulf brine and a 10 times diluted sample were used as the high-salinity and low-salinity brines, respectively. The hybrid method recovered more than 12% extra oil by switching from LSW to polymer in the Berea sandstone and more than 29% for the Bentheimer sample. Continuing LSW as the tertiary method and then converting to polymer flooding were also recommended based on their results [26]. Likewise, Tahir et al. showed that injection of polymer after LSW or smart water provides higher oil recovery than the oil recovery obtained by polymer injection before LSW. It is speculated that in this flooding sequence (i.e., polymer flooding before LSW), the LSW may follow the same paths as the polymer fluid, which inhibits the direct contact of LSW with the oil/rock interface. In contrast, the preinjection of LSW can alter the wettability of the rock by direct contact making oil detachment from rock surfaces easier, which aids the displacement of oil in the subsequent polymer flooding stage [27].

In addition to experimental studies, modeling approaches have also confirmed the benefits of hybrid LSW/polymer flooding. Khorsandi et al. [28] developed the first analytical solution for combined LSW/polymer flooding in sandstone to describe the synergy of this hybrid process, which allows recognizing the effective parameters and the mechanisms controlling oil recovery. This hybrid method was also simulated by Mohammadi and Jerauld [29], who showed higher oil recovery as displayed in **Figure 10** that shows that injection of polymer after LSW gives better performance than polymer injection and stand-alone LSW flooding. Furthermore, this study showed that wettability alteration by LSW and the simultaneous increase in brine viscosity, reduction of the relative brine permeability, and mobility control during polymer flooding improved the fractional flow of the process, as shown in **Figure 11**. The adjustment in fractional flow was also modeled by [30], which confirms the stable shock fronts during LSW/polymer flooding. As polymer cannot invade the inaccessible pore volume, the water remaining after LSW will be immobile in these pores, which reduces the channeling in the formation [31]. Hence, the hybrid method can control the unstable front and the channeling of low-viscosity LSW into the viscous oil. This makes the LSW/polymer hybrid method even more effective in heavy oil formations.

Other injection schemes have also been discussed and studied as reported in the literature. For example, Lee et al. modeled the process of polymer-assisted carbonated LSW flooding (PCLSWF) as a new hybrid method [32]. Different mechanisms are involved during PCLSWF, such as wettability modification by LSW, oil swelling, oil viscosity reduction by the gas, and mobility enhancement by the polymer. Also, higher pressure in the porous media in the presence of polymer leads to more dissolution of CO_2 in the brine and more transport of gas to the oil phase. Hence,

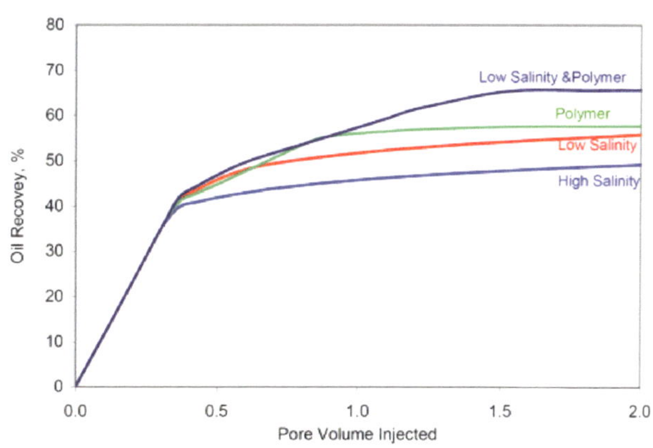

Figure 10.
Oil recovery modeling by hybrid method injection and comparison to other EOR approaches [29].

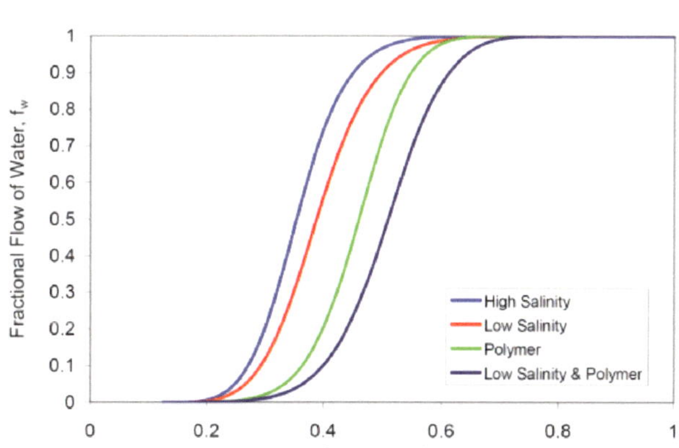

Figure 11.
Comparison of fractional flow of high-salinity, low-salinity, polymer, and hybrid flooding by modeling [29].

PCLSWF showed better performance than LSW flooding, hybrid low-salinity polymer flooding (LSPF), and hybrid carbonated LSW flooding (CLSWF), as shown in **Figure 12** [32]. Another study was conducted by Eikrem to analyze oil recovery by combining low-salinity injection and surfactant/polymer (SP) flooding. Cores with different initial wettability were flooded initially by the high-salinity water (HSW) and then with a surfactant solution in tertiary mode, followed by a polymer injection for mobility control. They found that injecting a 600 ppm HPAM polymer solution after the surfactant injection improved the ultimate recovery of oil [33]. Mjøs observed a similar behavior [34].

Interactions between polymer and LSW affect different governing parameters in the EOR process. Lower salinity has an influence on the polymer injectivity, retention, polymer stability, and rheological factors such as viscosity. Alsofi et al. studied the effect of LSW on the polymer solution properties by single-phase core flooding tests in carbonate samples. Anionic sulfonated polyacrylamide polymer was used as the polymer, and two samples of water with salinity of 244,000 and 24,400 mg/L were used as high- and low-salinity brines, respectively. They found that at lower-salinity concentration, coiling of the polymer backbone is reduced due to lower existing ions in LSW. Therefore, there are more repulsive interactions among the polymer chains causing the expansion of the polymer backbone, which results in higher viscosity of the polymer solution that leads to pressure buildup and slightly lower polymer injectivity [35].

The better solubility of polymer in LSW leads to the alteration of the polymer retention, which is critical for the technical and economical design of the process. In [35], it was indicated that the polymer retention decreased by 10–28%, which was noticeable. Expansion of the polymer chains due to repulsion results in the fitting of fewer polymer molecules on the adsorption rock surface sites, which reduces the retention. In [26], Almansour et al. also confirmed a reduction in the retention of polymer by using LSW. A decrease in polymer adsorption was also reported by [36]. Modeling has also showed that lower concentrations of polymer are required to establish a stable displacement front in the LSW/polymer approach than the HSW/polymer [29].

This synergistic effect also reduces the consumption of polymer, which is a positive point. At lower salinity, a lower polymer concentration is required to achieve the target viscosity, which makes the polymer flooding process more cost-competitive. In addition to the reduction of transportation costs, storage, and polymer handling, this behavior was also confirmed by [22]. Brine salinity also affects the stability of the polymer solution in some special conditions. Levitt et al. [37] showed that at lower ion composition, especially at low calcium concentration, HPAM is more stable at higher temperatures.

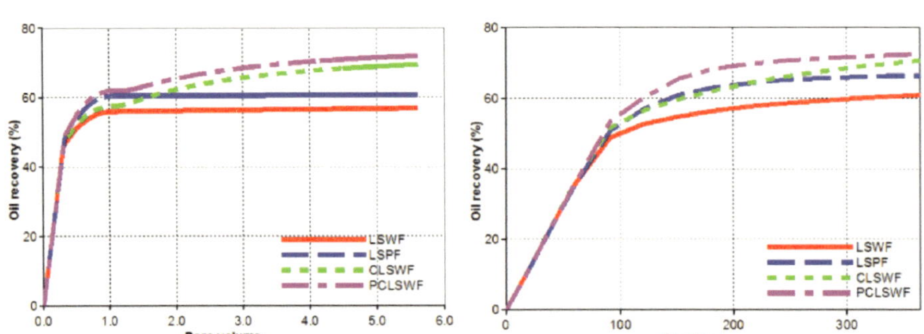

Figure 12.
Oil recovery of PCLSWF compared to other EOR methods in core scale (left) and pilot scale (right). LSWF is LSW flooding, LSPF is low-salinity polymer flooding, and CLSWF shows the carbonated LSW flooding [32].

Different operational parameters affect the performance of the LSW/polymer hybrid method, which should be considered at the design stage of the flooding process. The main parameters that should be studied are the initial wettability of the rock, smart water design (salinity concentration and composition), and the hybrid method initiation time. Shiran et al. studied the synergy of LSW/polymer flooding by core flooding experiments in sandstone samples [31]. The LSW was obtained by diluting 10 times seawater. Flopaam 3630S polyacrylamide with a hydrolysis degree of 25–30% was added to the LSW in concentrations of 300 and 1000 ppm to prepare the polymer solution. Aluminum citrate was added to the polymer solution to cross-link the polymer chains. This study revealed that the initial wettability of the porous medium was a critical factor affecting the success of the hybrid method, because incremental oil recovery was not observed during the polymer flooding step in the hybrid LSW/polymer injection scheme in core plug samples that were strongly water wet. However, the injection of polymer after LSW was effective for intermediate water-wet core samples, as shown in **Figure 13**. In water-wet formations, there are more adsorption sites available on the rock surfaces; thus, polymer retention is higher, which affects the performance of the method. Moreover, it was observed by [34] that the performance of the hybrid method was better in less water-wet conditions.

Another important parameter in the design of the hybrid method is the appropriate time to switch from conventional waterflooding to the hybrid method. A numerical simulation study of the hybrid LSW/polymer flooding conducted in 1D, 2D, and 3D reservoirs indicated that the hybrid method should be started at water cuts less than or equal to 75% to achieve improvement in oil recovery [30]. Hence, a comprehensive economic study is required to design the optimum case for field applications.

Ion management is a critical parameter to design smart waters and improve the performance of the diluted LSW/polymer flooding. Alteration in ion concentrations and ion types can affect the properties of the polymer solution, which must be considered during the design of hybrid polymer and smart water flooding. Experiments in [27] aimed to study the effect of active ions such as sulfate on the performance of the hybrid methods. The presence of sulfate affects the polymer solution viscosity, as shown in **Figure 14**.

The application of the LSW/polymer flooding offers economical and technical benefits for EOR applications. For example, the water cut is reduced compared to the stand-alone polymer or LSW flooding, as confirmed by simulations conducted

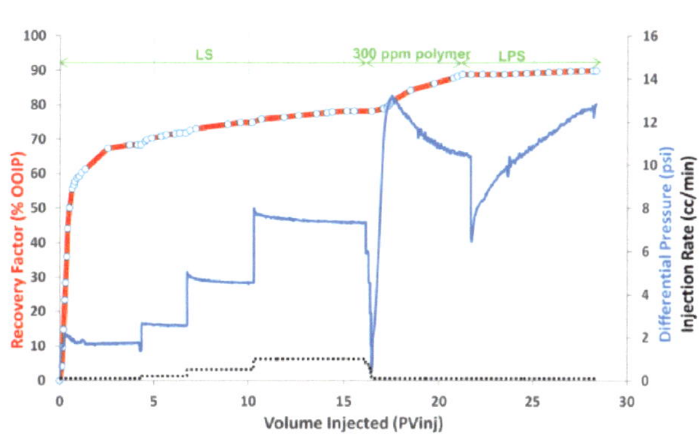

Figure 13.
Oil recovery during secondary mode LSW flooding followed by polymer flooding and linked polymer solution (LPS) flooding [31].

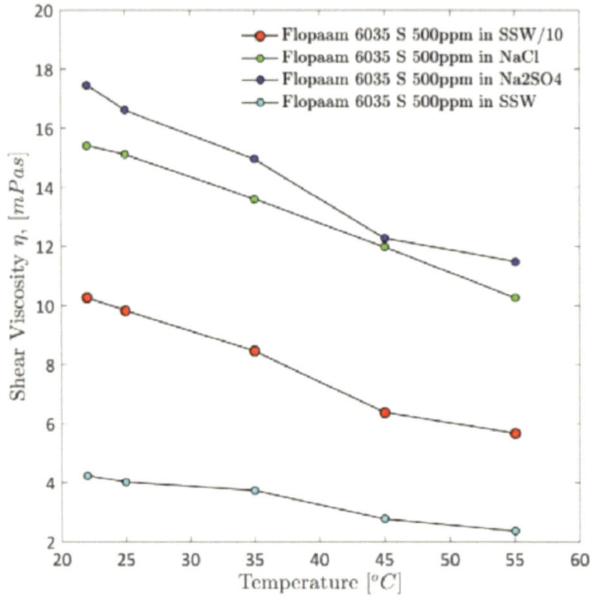

Figure 14.
Polymer steady-state viscosity at different temperatures for different types of solution brines (SSW stands for synthetic seawater) [27].

by Santo and Muggeridge [30]. Also, the oil recovery by LSW/polymer flooding was established to be slightly better than the conventional surfactant/polymer chemical flooding [22]. This is an important observation because the hybrid method could provide higher oil recovery at a lower cost. The studies reviewed in this section confirm the synergistic effect of combining LSW/polymer flooding.

4. LSW/surfactant hybrid EOR technique

The main mechanism responsible for the effectiveness of LSW in improving oil recovery is the alteration of the wettability of the rock toward more water wet that causes the detachment of oil films from the rock surface. The injection of surfactant reduces the interfacial tension (IFT) between crude oil and brine and alters the wettability of the rock reducing the capillary forces that have trapped oil in the porous media. Therefore, the combination of LSW and low-salinity surfactant (LSS) in LSW flooding could be an efficient approach by combining the effect of oil layer destabilization by LSW and reduction of the IFT by the surfactant. This hybrid method provides higher incremental oil recovery than either stand-alone techniques.

LSW makes the environment more favorable for an effective surfactant flooding, while LSS solubilizes some of the residual oil via Winsor type II microemulsion. Several studies have reported high tertiary oil recovery values by surfactant injection after LSW flooding in both carbonate and sandstone formations. For example, according to [38], 5–7% incremental oil recovery was observed by injection of sodium dodecyl benzene sulfonate surfactant (SDBS) after LSW into sandstones. Similarly, Alameri et al. observed up to 10% incremental oil recovery by LSS after LSW injection into carbonate core samples [39].

Application of the LSW/LSS hybrid method results in lower surfactant consumption, lower operational costs, and fewer operational problems. For instance, it is less challenging to achieve a low IFT during surfactant flooding at low-salinity conditions. Likewise, in these conditions, there is reduced surfactant retention and increased surfactant stability and solubility.

In this hybrid process, the dominant mechanisms for increased oil recovery are wettability alteration by LSW and IFT reduction between the crude oil and brine by LSS. Alagic et al. studied this hybrid method through core flooding tests and analyzed the performance of LSS injection after LSW into sandstone samples. Olefin sulfonate was used as the surfactant in these tests. The results show more than 90% oil recovery of the OOIP when surfactant is injected after LSW. This injection sequence produces higher oil recovery than the injection of surfactant after high-salinity water (74% OOIP), as shown in **Figure 15** [40]. Injection of LSW makes the system more water wet, which aids the detachment of oil droplets and reduces the capillary force, making easier the displacement of oil during the surfactant flooding process. Surfactant injection in a low-salinity environment is more effective, as the presence of the divalent ions contained in high-salinity water attenuates IFT reduction due to the formation of water-in-oil microemulsions.

The achievement of ultralow IFT during LSS is critical; however, several studies have demonstrated that wettability alteration by the hybrid LSW/LSS approach can be very effective. Therefore, in this process, wettability alteration is considered the dominant mechanism for oil recovery. Johannessen et al. [41] used branched C12–13 alcohol-xPO-sulfates as the surfactant, olefin sulfonate as the co-surfactant, and secondary butanol (SBA) as the cosolvent to evaluate the performance of LSW/LSS flooding from Berea sandstone cores in terms of oil recovery. Two flooding conditions were studied: in the first case, the core was flooded using low-salinity brine corresponding to 0.07 times the seawater salinity (diluted case), and the other one was flooded at the optimum salinity of surfactant flooding obtained from phase behavior screening and IFT measurements (optimum salinity case). The same oil recovery was observed for both cases, as shown in **Figure 16**. Incremental oil recovery beyond expectations by the capillary number changes is explained by the synergistic effect of the low-salinity surfactant flooding, even at IFT values above ultralow IFT values.

The same behavior was observed by Khanamiri et al. during LSW/LSS flooding of sandstone samples [42]. The incremental oil recovery by LSS after LSW was in the range of 2–6% OOIP. They observed that the oil mobilization in the process was mostly due to wettability alteration by LSW and LSS and IFT reduction cannot be considered as the dominant mechanism. As the salinity of the solution was not at the optimum salinity condition, ultralow IFT was not achieved; however, the significant wettability alteration caused by LSS flooding verified from contact angle measurements (quartz crystal microbalance (QCM) on silica-coated crystals) compensated the effect of having a value of IFT higher than the desired ultralow IFT

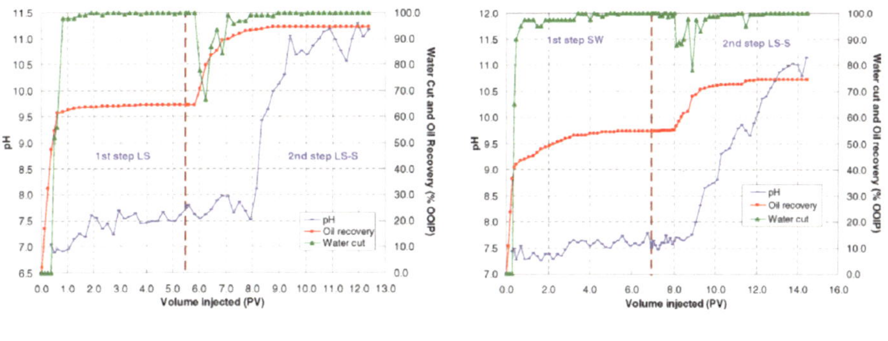

Figure 15.
Oil recovery, water cut, and effluent pH during LSW/LSS injection (left) and synthetic seawater (SSW)/LSS flooding (right) [40].

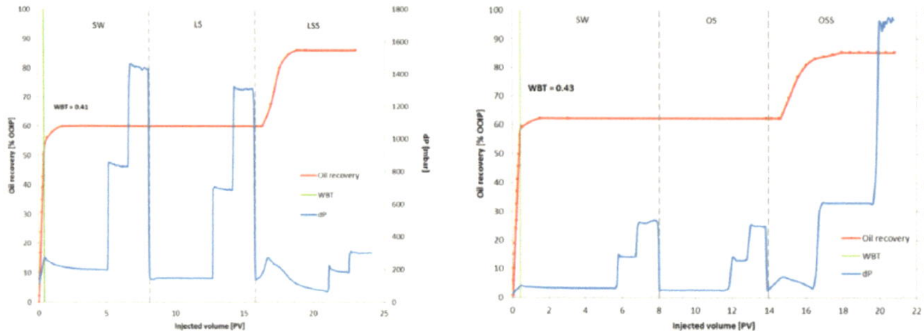

Figure 16.
Oil recovery and dP profiles for injection of seawater followed by LSW and low-salinity surfactant at diluted case (left) and optimum salinity (right) [41].

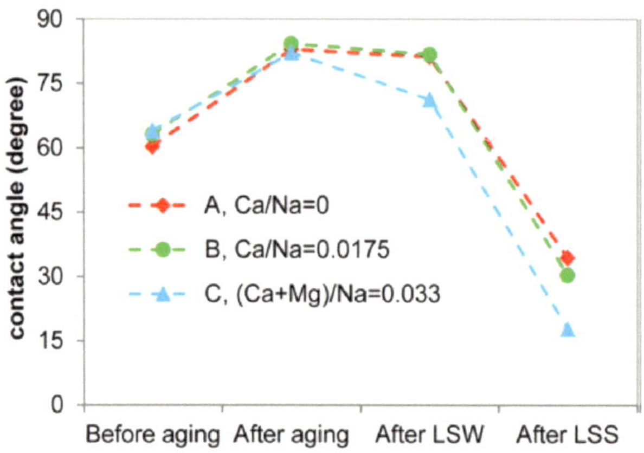

Figure 17.
Contact angle for deionized water droplet/air/silica after different treatments [42].

value. **Figure 17** shows the alteration in contact angle for different treated brines. This figure indicates the initial contact angle and the angles after aging in high-salinity water, in LSW, and in LSS. As can be seen, wettability alteration toward water-wet condition occurred for all cases with different types of LSW. However, different ion contents affected the magnitude of wettability alteration.

The effect of surfactant on wettability alteration was also observed by Teklu et al. [43]. They measured the contact angle of different carbonate and sandstone rock disks saturated with oil in low-salinity brine in the presence and absence of nonionic ethoxylated alcohol surfactant. They found that the presence of surfactant decreases the contact angle and makes the system more water wet. Reduction in IFT and alteration of rock wettability by LSS can increase oil recovery in cases when LSW is not effective alone. For example, core flooding experiments conducted by Spildo et al. showed that the application of surfactant-free LSW does not increase oil recovery, while LSS produces incremental oil recovery, as shown in **Figure 18** [44].

These synergistic mechanisms (i.e., wettability alteration and reduction of IFT) in the hybrid LSW/LSS method provide higher oil recovery than the recovery expected from the capillary desaturation curves (CDC). Studies in [44, 45] showed that during the LSS after LSW, the capillary number was about 10^{-4}, which is not high enough to achieve noticeable incremental oil recovery, as observed in the experiments [46]. Hence, oil detachment and redistribution due to rock wettability alteration during LSW make the LSS performance better than the estimation from the CDC.

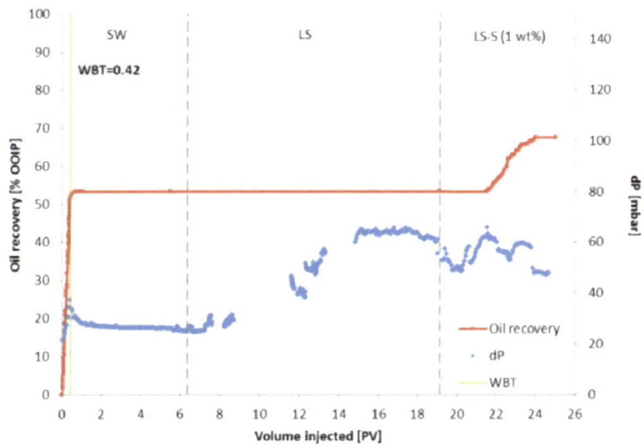

Figure 18.
Oil recovery and pressure drop profile as a function of pore volume injected into sandstone core during SW/LSW/LSS injection [44].

Shaddel et al. evaluated the incremental oil recovery obtained from LSW, LSW/surfactant, and LSW/alkali injection in Berea and Bentheimer sandstones in tertiary mode. Sodium hydroxide and sodium dodecyl sulfate dissolved in 0.01 LSW were used as alkali and surfactant solutions, respectively. The authors considered that the LSW/alkali is a convenient process as a hybrid method due to the lower operation costs [47].

Brine composition is an important variable during the design of surfactant flooding. The retention (i.e., adsorption) of surfactant molecules onto porous media is considered a critical issue; thus, the reduction of surfactant adsorption onto the porous media enhances the quality of a surfactant flooding project from the technical and economical points of view. Glover et al. reported that low-salinity brine surfactant adsorption is reduced [48]. Lower surfactant retention at lower salinities was also observed in [43, 45]. Additionally, Johannessen et al. [41] observed that the surfactant retention values were lower at very-low-salinity brines than at the optimum salinity condition, as measured by retention tests, which are shown in **Figure 19**. The greater area under the production curve for the LSS condition corresponds to lower surfactant retention. This implies that the hybrid LSW/LSS method is more economically efficient than the injection of surfactant at the ultralow IFT formulation. Tests performed by Araz and Kamyabi showed that precipitation of SDBS surfactant in LSW occurs when salinity concentration is above 1000 ppm [49].

The type and concentration of divalent cations also influence the performance of surfactant flooding. For example, Enge [38] showed that the content of divalent ions in brines affects the precipitation of surfactant. Calcium cations affect the behavior of surfactant adsorption and surfactant precipitation due to interactions with calcium (negative effect) and the stabilization of micelles (positive effect). Hence, there is a limit for divalent cation concentration which must be considered during the LSS design stage. Other properties of surfactant solutions are affected in a low-salinity environment, such as solubility and retention. For instance, Alagic et al. observed that the surfactant solubility is improved in low-salinity brine, especially in the absence of divalent cations [45].

The initial wettability of the porous media is a key parameter in the success of most hybrid LSW EOR methods. Alagic et al. [45] studied the effect of crude oil aging on LSW/LSS flooding in sandstones. They used a sulfonate surfactant added to the low-salinity brine in the LSS injection period. The performance of the LSS

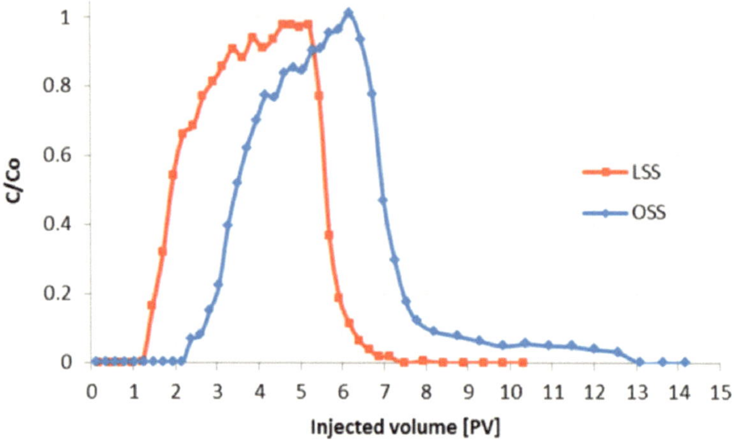

Figure 19.
Normalized produced surfactant concentration for surfactant in very LSW (LSS) and in water with the optimum salinity (OSS) [41].

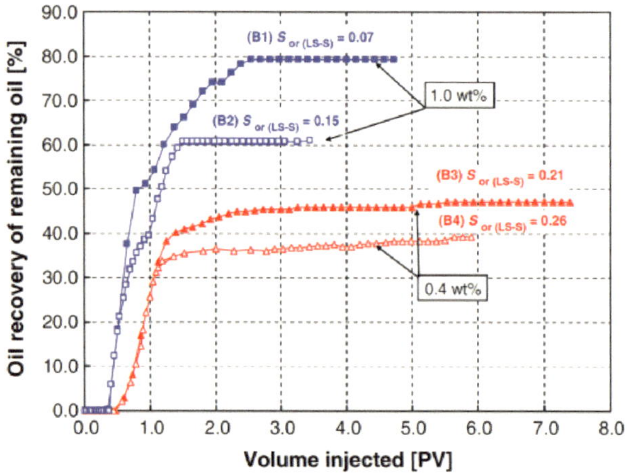

Figure 20.
Oil recovery of remaining oil (after LSW flooding) in aged (B1, B3) and unaged (B2, B4) sandstone samples for two surfactant concentrations (1 and 0.4 wt%) [45].

was observed to be better in cores aged by oil. In addition, LSS recovered more oil at a higher concentration of surfactant. **Figure 20** shows the performance of oil recovery in the LSS stage for aged and unaged samples. The same trend was also observed in [33, 34]. Another variable affecting the performance of hybrid methods is the composition of the LSW. Araz and Kamyabi examined the effect of the LSW composition on the performance of core flooding by LSW/LSS in sandstones. They found that alteration in the composition of different ions in LSW and LSS affected oil recovery [49]. Therefore, it is essential to study the effect of ion composition in the LSW/LSS process in terms of oil recovery.

The stability of the displacement front in the LSW/LSS process is a critical issue to achieve a successful recovery. Tavassoli et al. [50] showed that unstable fronts of surfactant floods due to the high velocity result in slow oil recovery. This problem can be solved through the combination of LSW with surfactant/polymer (SP). In this process, three different oil recovery mechanisms are active such as wettability alteration, reduction in IFT, and mobility control. In [51], Wang et al. studied the hybrid low-salinity surfactant/polymer flooding as an EOR method in carbonates.

They observed more recovery by LSW/SP in LSW than HSW/SP. This study confirmed the destabilization of oil layers after LSW injection prior to SP flooding by pressure drop analysis. The pressure drop was more significant due to the formation of an oil bank after LSW injection.

Most of the research conducted in the area of hybrid LSW/LSS or LSW/SP processes has been based on experimental studies. There are few simulation studies published on the effectiveness of the hybrid LSW/LSS method. The lack of modeling and optimization of the process through simulation studies is obvious in this field. In [50], Tavassoli et al. applied UTCHEM-IPhreeqc to model LSW as a function of geochemical reactions and surfactant flooding. Their simulations were in good agreement with experiments carried out by [27]. This study demonstrated the importance of the surfactant selection, injection sequences, and operational parameters such as brine salinity and surfactant solution injection rate to achieve incremental oil recovery. Therefore, more modeling studies are justified.

5. LSW/nanofluid hybrid EOR technique

Nanoparticles are used in EOR processes due to their size and high surface area; thus, nanoparticles flow without difficulty through the pore/throat network in porous media. Nanoparticles enhance oil recovery by the following mechanisms: IFT reduction, wettability alteration, improvement of mobility ratio, and in situ emulsification [52, 53]. For example, SiO_2 particles are hydrophilic and can be injected into the porous media to alter the rock wettability toward more water wet [54, 55].

Despite the positive effects of LSW on oil recovery, LSW flooding changes the chemical environment (pH, ionic strength, and temperature) in the porous media, which may lead to the detachment of reservoir particles. As the salinity of the injected brine becomes less than the critical salinity concentration, fine migration initiates, which leads to formation damage [56, 57]. Fine migration can enhance oil recovery by mobility control through the blockage of high-permeability layers, but fine migration can also cause severe damage to the near-wellbore zone. Therefore, it is desirable to control fine migration and to take advantage of its positive effects far from the wellbore to minimize its damaging effects near the wellbore. Some researchers have stated that combining LSW and NPs may help to overcome the detrimental effects of formation damage associated with low-salinity flooding [58, 59]. Hence, nanoparticles in nanofluid form can be used as a hybrid approach with LSW flooding to improve the performance of this method.

Nanoparticles enhance the attractive forces between fine particles and grain surfaces, particularly by changing the surface zeta potentials of fine particles [60]. Nanofluid pretreatment prior to the injection of LSW can reduce the side effects of fine migration by decreasing the injection pressure drop [53]. Abhishek et al. [52] studied nanoparticle adsorption at different salinities and observed that during hybrid LSW and nanofluid injection, the adsorption of nanoparticles prevents fine migration.

Another approach is to apply the hybrid nanofluid/LSW method to alter wettability and interfacial tension. Hydrophilic nanoparticles adsorb on the rock surface, and water molecules accumulate around them. This changes the wettability of the rock to be more water wet and improves oil recovery. Sadatshojaei et al. [60] coupled silica nanoparticles with LSW prepared by dilution of Persian Gulf brine. In this study, IFT, wettability, and zeta potential were measured to analyze the carbonate rock-low-salinity fluid interaction in the presence of nanofluid. It was observed that the influence of the nanofluid on the wettability of the porous media was dominant relative to the IFT reduction; thus, wettability alteration can be considered as the main mechanism for oil recovery. Moreover, Ding et al. [61] introduced nanoparticle-assisted low-salinity hot water (LSHW) injection for heavy oil recovery. Flooding tests were conducted on silica sand packs saturated with heavy oil to compare the effect of LSW flooding, 0.05 wt% SiO_2 nanoparticle-dispersed LSW flooding, and 0.05 wt% Al_2O_3 nanoparticle-dispersed LSW flooding on oil recovery. They observed higher oil recovery by the injection of NP/LSW than LSW alone. Also, they found that Al_2O_3 NPs were more effective in recovering oil due to the greater reduction in IFT.

Generally, the average nanoparticle size, specific surface area, and stability of nanofluids are important in the performance of this hybrid method. The issue of nanofluid stability at low-salinity conditions is challenging and should be considered in the application of this hybrid method. The range of stability constraints (such as zeta potential) of colloidal systems of nanofluids coupled with ions (i.e., LSW) is typically wider so that in the presence of both elements (nanoparticles and ions), longer-lasting stable solutions can be attained [60].

6. LSW/hot water hybrid EOR technique

Heavy oil is conventionally recovered by thermal-based approaches. Thermal energy in combination with LSW water flooding, in the form of hot LSW water flooding, can be applied to simultaneously decrease the viscosity of the heavy oil and alter the wettability of the porous media to attain higher oil recovery. Alotaibi et al. and Tang et al. reported lower contact angle and higher oil recovery, respectively, after increasing the displacement temperature in LSW injection [62, 63]. Contrary to this observation, Soraya et al. [64] found lower oil recovery at higher temperatures by tertiary LSW. However, another study [65] demonstrated that the injection of hot LSW yielded significant incremental oil recovery. In this work, hot LSW at a concentration of 200 ppm salinity was injected after hot HSW at a concentration of 15,000 ppm salinity, which yielded about 25% OOIP incremental oil recovery, as shown in **Figure 21**. In addition, in this study the injection of steam is proposed after hot LSW flooding to further enhance the oil recovery.

A similar trend was reported by Ding et al. by injecting nanoparticle-assisted low-salinity hot water (LSHW) into silica sand packs to recover heavy oil. Temperatures of 17, 45, and 70°C were applied under different scenarios of high- and low-salinity brines and nanofluids. LSW was found to provide better

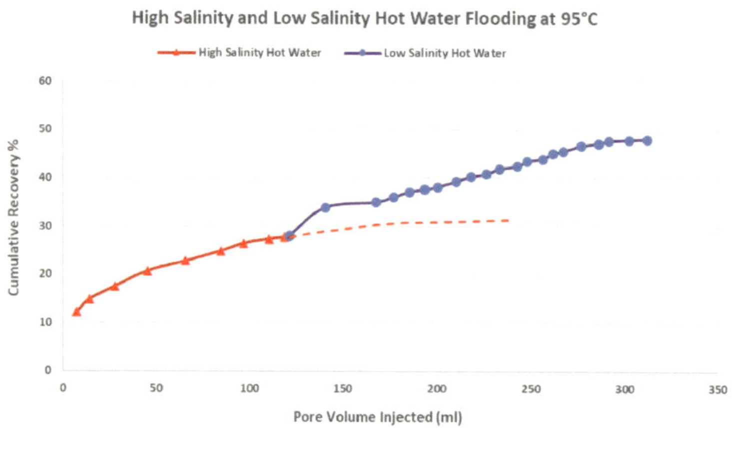

Figure 21.
Oil recovery by hot HSW and hot LSW [65].

performance than the HSW under ambient temperature. Moreover, in all cases in the presence or absence of nanoparticles, increasing the displacement temperature yielded higher ultimate oil recoveries up to 23%. Increased temperature also restrains the growth rate of water cut [65].

The results to date in terms of enhanced oil recovery for all hybrid LSW-based EOR methods are promising. So far, experiments at laboratory scale have been carried out, while modeling studies and pilot field applications are scarce. Numerical simulation studies are necessary to provide more practical insights on the effectiveness of hybrid methods especially important for field implementations.

7. Conclusion

The main idea of this chapter is to demonstrate the synergistic EOR effects of combining chemical/gas-based/thermal methods with low-salinity water and the related underlying mechanisms in both sandstone and carbonate rocks. Hybrid EOR methods are utilized to bypass or improve operational, environmental, and economical shortcomings of individually implemented methods. Many experimental and modeling studies have confirmed this potential synergy by mentioning wettability alteration toward more water-wet condition as the main mechanism of LSW method.

Lower ion concentration of LSW allows gas (typically CO_2) molecules to dissolve in water phase in higher extent which results in gas/oil contact and improved front stability. However, some studies cast doubt on this idea as the more gas dissolves in water, the lower free gas is available to decrease oil viscosity and ultimately improve oil mobility.

Surfactants are known as the agents which are utilized to decrease the IFT, consequently capillary forces, between oil and water in order to enhance microscopic sweep efficiency. LSW provides more detached oil droplets to be produced due to lowered capillary forces by surfactants.

Polymers cause higher oil recovery by increasing water viscosity which lowers mobility ratio after reservoir fluid redistribution by LSW injection due to wettability alteration.

Thermal methods are applied in heavy oil reservoirs to help oil recovery under the mechanism of improved oil mobility, which can be more advantageous if higher detached oil droplets are provided by LSW injection. Lastly, nanoparticles have been introduced to be beneficial as they can improve wettability alteration process by LSW injection.

Despite the extensive promising works mentioned in this chapter, many contradictions, ambiguities, and unexamined issues, especially in carbonate rocks, require more investigation in LSW hybrid methods regarding underlying mechanisms, field implementation viability, operational considerations, and economical feasibility by both experimental and modeling assessments.

Author details

Peyman Pourafshary* and Nikoo Moradpour
School of Mining and Geosciences, Nazarbayev University, Nur-Sultan, Kazakhstan

*Address all correspondence to: peyman.pourafshary@nu.edu.kz

References

[1] Teklu TW, Alameri W, Graves RM, Kazemi H, AlSumaiti AM. Low-salinity water-alternating-CO_2EOR. Journal of Petroleum Science and Engineering. 2016;**142**:101-118. DOI: 10.1016/j.petrol.2016.01.031

[2] Li Y-K, Nghiem LX. Phase equilibria of oil, gas and water/brine mixtures. Canadian Journal of Chemical Engineering. 1986;**64**:486-496. DOI: 10.1002/cjce.5450640319

[3] Pollack NR, Enick RM, Mangone DJ, Morsi BI. Effect of an aqueous phase on CO_2/tetradecane and CO_2/maljamar-crude-oil systems. SPE Journal. 1988;**3**:533-541. DOI: 10.2118/15400-PA

[4] Enick RM, Klara SM. CO_2 solubility in water and brine under reservoir conditions. Chemical Engineering Communications. 1990;**90**:23-33. DOI: 10.1080/00986449008940574

[5] Duan Z, Sun R. An improved model calculating CO_2 solubility in pure water and aqueous NaCl solutions from 273 to 533 K and from 0 to 2000 bar. Chemical Geology. 2003;**193**:257-271. DOI: 10.1016/S0009-2541(02)00263-2

[6] Aleidan AA, Mamora DD. SWACO_2 and WACO_2 efficiency improvement in carbonate cores by lowering water salinity. Canadian Unconventional Resources and International Petroleum Conference. Jan 2010. DOI: 10.2118/137548-MS

[7] Jiang H, Nuryaningsih L, Adidharma H. The effect of salinity of injection brine on water alternating gas performance in tertiary miscible carbon dioxide flooding: Experimental study. SPE Western Regional Meeting. 2010:1984. DOI: 10.2118/132369-MS

[8] Kumar HT, Shehata AM, Nasr-El-Din HA. Effectiveness of low-salinity and CO_2 flooding hybrid approaches in low-permeability sandstone reservoirs. SPE Trinidad and Tobago Section Energy Resources Conference. 2016. DOI: 10.2118/180875-MS

[9] Al-Shalabi EW, Sepehrnoori K, Pope G. Numerical modeling of combined low salinity water and carbon dioxide in carbonate cores. Journal of Petroleum Science and Engineering. 2016;**137**:157-171. DOI: 10.1016/j.petrol.2015.11.021

[10] Sheng J. Enhanced Oil Recovery Field Case Studies. Waltham, MA: Gulf Professional Publishing; 2013. p. 712

[11] Ramanathan R, Shehata AM, Texas A. Water alternating CO_2 injection process—Does modifying the salinity of injected brine improve oil recovery? OTC Brazil. Offshore Technology Conference. 2015. DOI: 10.4043/26253-MS

[12] Ramanathan R, Shehata AM, Nasr-El-Din HA. Effect of rock aging on oil recovery during water-alternating-CO_2 injection process: An interfacial tension, contact angle, coreflood, and CT scan study. SPE Improved Oil Recovery Conference. 2016:1-18. DOI: 10.2118/179674-MS

[13] Al-Abri H, Pourafshary P, Mosavat N, Al Hadhrami H. A study of the performance of the LSWA CO_2EOR technique on improvement of oil recovery in sandstones. Petroleum. 2019;**5**(1):58-66. DOI: 10.1016/j.petlm.2018.07.003

[14] AlQuraishi AA, Amao AM, Al-Zahrani NI, AlQarni MT, AlShamrani SA. Low salinity water and CO_2 miscible flooding in Berea and Bentheimer sandstones. Journal of King Saud University—Engineering Sciences. 2019;**31**(3):286-295. DOI: 10.1016/J.JKSUES.2017.04.001

[15] Zolfaghari H, Zebarjadi A, Shahrokhi O, Ghazanfari MH. An experimental study of CO_2-low salinity water alternating gas injection in sandstone heavy oil reservoirs. Iranian Journal of Oil & Gas Science and Technology. 2013;**2**:37-47. DOI: 10.22050/IJOGST.2013.3643

[16] Yang D, Tontiwachwuthikul P, Gu Y. Interfacial tensions of the crude oil + reservoir brine + CO_2 systems at pressures up to 31 MPa and temperatures of 27°C and 58°C. Journal of Chemical & Engineering Data. 2005;**50**:1242-1249. DOI: 10.1021/je0500227

[17] Bennion DB, Bachu S. Correlations for the interfacial tension between supercritical phase CO_2 and equilibrium brines at in situ conditions. SPE Journal. 2008. DOI: 10.2118/114479-MS

[18] Kulkarni MM, Rao DN. Experimental investigation of miscible and immiscible water-alternating-gas (WAG) process performance. Journal of Petroleum Science and Engineering. 2005;**48**(1-2):1-20. DOI: 10.1016/j.petrol.2005.05.001

[19] Al-Shalabi EW, Sepehrnoori K, Pope G. Geochemical investigation of the combined effect of injecting low salinity water and carbon dioxide on carbonate reservoirs. Energy Procedia. 2014;**63**:7663-7676. DOI: 10.1016/j.egypro.2014.11.800

[20] Dang CTQ, Nghiem LX, Chen Z. CO_2 low salinity water alternating gas: A new promising approach for enhanced oil recovery. SPE Improved Oil Recovery Symposium. 2014:1-19. DOI: 10.2118/169071-MS

[21] Piñerez Torrijos ID, Puntervold T, Strand S, Austad T, Bleivik TH, Abdullah HI. An experimental study of the low salinity smart water—Polymer hybrid EOR effect in sandstone material. Journal of Petroleum Science and Engineering. 2018;**164**:219-229. DOI: 10.1016/j.petrol.2018.01.031

[22] Al Sofi AM, Wang J, Al Boqmi AM, Al Otaibi MB, Ayirala SC, Al Yousef AA. Smart water synergy with chemical EOR for a slightly viscous Arabian heavy reservoir. SPE Heavy Oil Conference and Exhibition. 2016. DOI: 10.2118/184163-MS

[23] Chie Kozaki: Efficiency of Low Salinity Polymer Flooding in Sandstone Cores. The University of Texas at Austin; 2012

[24] Vermolen ECM, Pingo Almada M, Wassing BM, Ligthelm DJ, Masalmeh SK. Low-salinity polymer flooding: Improving polymer flooding technical feasibility and economics by using low-salinity make-up brine. International Petroleum Technology Conference. Jan 2014. DOI: 10.2523/IPTC-17342-MS

[25] Borazjani S, Bedrikovetsky P, Farajzadeh R. Analytical solutions of oil displacement by a polymer slug with varying salinity. Journal of Petroleum Science and Engineering. 2016;**140**: 28-40. DOI: 10.1016/j.petrol.2016.01.001

[26] Almansour AO, AlQuraishi AA, AlHussinan SN, AlYami HQ. Efficiency of enhanced oil recovery using polymer-augmented low salinity flooding. Journal of Petroleum Exploration and Production Technologies. 2017;7:1149-1158. DOI: 10.1007/s13202-017-0331-5

[27] Tahir M, Hincapie RE, Foedisch H, Abdullah H, Ganzer L. Impact of sulphates presence during application of smart water flooding combined with polymer flooding. SPE Europec featured at 80th EAGE Conference and Exhibition. 2018. DOI: 10.2118/190796-MS

[28] Khorsandi S, Qiao C, Johns RT. Displacement efficiency for low-salinity

polymer flooding including wettability alteration. SPE Journal. 2017;**22**:417-430. DOI: 10.2118/179695-PA

[29] Mohammadi H, Jerauld G. Mechanistic modeling of the benefit of combining polymer with low salinity water for enhanced oil recovery. SPE Improved Oil Recovery Symposium. 2012. DOI: 10.2118/153161-MS

[30] Santo A, Muggeridge A. An investigation into the benefits of combined polymer-low salinity water flooding. SPE Asia Pacific Oil and Gas Conference and Exhibition. 2018:1-19. DOI: 10.2118/192111-MS

[31] Shaker Shiran B, Skauge A. Enhanced oil recovery (EOR) by combined low salinity water/polymer flooding. Energy & Fuels. 2013;**27**:1223-1235. DOI: 10.1021/ef301538e

[32] Lee JH, Kim GW, Lee KS. Evaluation of hybrid EOR as polymer-assisted carbonated waterflood in calcite cemented sandstone reservoir. Offshore Technology Conference. 2017:1-19. DOI: 10.4043/28125-MS

[33] Eikrem OS. Low Salinity Waterflood in Combination with Surfactant/Polymer; Effects of Kinetics and Brine Composition. Norway: University of Bergen; 2014

[34] Mjøs JES. Low Salinity Waterflood in Combination with Surfactant/Polymer: Effect of Brine Composition. Norway: University of Bergen; 2014

[35] AlSofi AM, Wang J, Kaidar ZF. Smartwater synergy with chemical EOR: Effects on polymer injectivity, retention and acceleration. SPE Journal. 2018;**166**:274-282. DOI: 10.1016/j.petrol.2018.02.036

[36] Sorbie KS. In: Sorbie KS, editor. Polymer-Improved Oil Recovery. 1st ed. Glasgow: Boca Raton, Fla: CRC Press; 1991. p. 359

[37] Levitt D, Pope GA. Selection and screening of polymers for enhanced-oil recovery. SPE Symposium on Improved Oil Recovery. 2008:19-23. DOI: 10.2118/113845-MS

[38] Enge IB. The effect of brine composition and rock type on oil recovery by the use of combined low-salinity waterflooding and surfactant flooding. NTNU; 2014

[39] Alameri W, Teklu T, Graves R, Kazemi H, AlSumaiti A. Low-salinity water-alternate-surfactant in low-permeability carbonate reservoirs. IOR 2015-18th European Symposium on Improved Oil Recovery. 2015. DOI: 10.3997/2214-4609.201412158

[40] Alagic E, Skauge A. Combined low salinity brine injection and surfactant flooding in mixed-wet sandstone cores. Energy & Fuels. 2010;**24**:3551-3559. DOI: 10.1021/ef1000908

[41] Johannessen AM, Spildo K. Enhanced oil recovery (EOR) by combining surfactant with low salinity injection. Energy & Fuels. 2013;**27**:5738-5749. DOI: 10.1021/ef400596b

[42] Hosseinzade Khanamiri H, Baltzersen Enge I, Nourani M, Stensen JÅ, Torsæter O, Hadia N. EOR by low salinity water and surfactant at low concentration: Impact of injection and in situ brine composition. Energy & Fuels. 2016;**30**:2705-2713. DOI: 10.1021/acs.energyfuels.5b02899

[43] Teklu TW, Alameri W, Kazemi H, Graves RM, AlSumaiti AM. Low salinity water–surfactant–CO_2EOR. Petroleum. 2017;**3**:309-320. DOI: 10.1016/j.petlm.2017.03.003

[44] Spildo K, Johannessen A, Skauge A. Low salinity waterflood at

reduced capillarity. SPE Improved Oil Recovery Symposium. 2012:1-13. DOI: 10.2118/154236-MS

[45] Alagic E, Spildo K, Skauge A, Solbakken J. Effect of crude oil ageing on low salinity and low salinity surfactant flooding. Journal of Petroleum Science and Engineering. 2011;**78**:220-227. DOI: 10.1016/j.petrol.2011.06.021

[46] Garnes JM, Mathisen AM, Scheie A, Skauge A. Capillary number relations for some north, sea reservoir sandstones. SPE/DOE Enhanced Oil Recovery Symposium. 1990. DOI: 10.2118/20264-MS

[47] Shaddel S, Tabatabae-Nejad SA. Alkali/surfactant improved low-salinity waterflooding. Transport in Porous Media. 2015;**106**:621-642. DOI: 10.1007/s11242-014-0417-1

[48] Glover CJ, Puerto MC, Maerker JM, Sandvik EL. Surfactant phase behavior and retention in porous media. SPE Journal. 1979;**19**:183-193. DOI: 10.2118/7053-PA

[49] Araz A, Kamyabi F. Experimental study of combined low salinity and surfactant flooding effect on oil recovery

[50] Tavassoli S, Kazemi A, Korrani N, Pope GA. Low-salinity surfactant flooding—A multimechanistic enhanced-oil-recovery method. SPE Journal. 2016;**21**(3):744-760. DOI: 10.2118/173801-PA

[51] Wang J, Ayirala SC, AlSofi AM, Al-Yousef AA, Aramco S. Smartwater synergy with surfactant polymer flooding for efficient oil mobilization in carbonates. SPE EOR Conference at Oil and Gas West Asia. March 2018. DOI: 10.2118/190334-MS

[52] Abhishek R, Hamouda AA, Murzin I. Adsorption of silica nanoparticles and its synergistic effect on fluid/rock interactions during low salinity flooding in sandstones. Colloids and Surfaces A: Physicochemical and Engineering Aspects. 2018;**555**:397-406. DOI: 10.1016/j.colsurfa.2018.07.019

[53] Yuan B, Moghanloo RG, Zheng D. Enhanced oil recovery by combined nanofluid and low salinity water flooding in multi-layer heterogeneous reservoirs. SPE annual technical conference and exhibition. 2016. DOI: 10.2118/181392-MS

[54] Moradi B, Pourafshary P, Jalali F, Mohammadi M. Effects of nanoparticles on gas production, viscosity reduction, and foam formation during nanofluid alternating gas injection in low and high permeable carbonate reservoirs. Canadian Journal of Chemical Engineering. 2017;**95**:479-490. DOI: 10.1002/cjce.22699

[55] Moradi B, Pourafshary P, Jalali F, Mohammadi M, Emadi MA. Experimental study of water-based nanofluid alternating gas injection as a novel enhanced oil-recovery method in oil-wet carbonate reservoirs. Journal of Natural Gas Science and Engineering. 2015;**27**:64-73. DOI: 10.1016/j.jngse.2015.07.009

[56] Habibi A, Ahmadi M, Pourafshary P, Ayatollahi S. Fines migration control in sandstone formation by improving silica surface zeta potential using a nanoparticle coating process. Energy Sources, Part A: Recovery, Utilization, and Environmental Effects. 2014;**36**:2376-2382. DOI: 10.1080/15567036.2011.569836

[57] Ahmadi M, Habibi A, Pourafshary P, Ayatollahi S. Zeta-potential investigation and experimental study of nanoparticles deposited on rock surface to reduce fines migration. SPE Journal. 2013;**18**:534-544. DOI: 10.2118/142633-PA

[58] Assef Y, Arab D, Pourafshary P. Application of nanofluid to control fines

migration to improve the performance of low salinity water flooding and alkaline flooding. Journal of Petroleum Science and Engineering. 2014;**124**: 331-340. DOI: https://doi.org/10.1016/j.petrol.2014.09.023

[59] Assef Y, Pourafshary P, Hejazi H. Controlling interactions of colloidal particles and porous media during low salinity water flooding and alkaline flooding by MgO nanoparticles. SPE EOR Conference at Oil and Gas West Asia. March 2016. DOI: 10.2118/179768-MS

[60] Jamialahmadi M, Esmaeilzadeh F, Ghazanfari MH. Effects of low-salinity water coupled with silica nanoparticles on wettability alteration of dolomite at reservoir temperature AU—Sadatshojaei, Erfan. Petroleum Science and Technology. 2016;**34**:1345-1351. DOI: 10.1080/10916466.2016.1204316

[61] Ding Y, Zheng S, Meng X, Yang D. Low salinity hot water injection with addition of nanoparticles for enhancing heavy oil recovery under reservoir conditions. Journal of Energy Resources Technology. 2019;**141**(7):072904. DOI:10.1115/1.4042238

[62] Alotaibi MB, Azmy R, Nasr-El-Din HA. A comprehensive EOR study using low salinity water in sandstone reservoirs. SPE improved oil recovery symposium. 2010. DOI: 10.2118/129976-MS

[63] Tang GQ, Morrow NR. Salinity, temperature, oil composition, and oil recovery by waterflooding. SPE Journal. 1997;**12**:269-276. DOI: 10.2118/36680-PA

[64] Soraya B, Malick C, Philippe C, Bertin HJ, Hamon G. Oil recovery by low-salinity brine injection: Laboratory results on outcrop and reservoir cores. SPE Annual Technical Conference and Exhibition. 2009. DOI: 10.2118/124277-MS

[65] Abass E: Experimental investigation of low salinity hot water injection to enhance the recovery of heavy oil reservoirs. SPE Journal 2013;15-17. DOI: http://dx.doi.org/10.2118/164768-MS.

Chapter 2

Nanotechnology Application in Chemical Enhanced Oil Recovery: Current Opinion and Recent Advances

Afeez Gbadamosi, Radzuan Junin, Muhammad Manan, Augustine Agi and Jeffrey Oseh

Abstract

Chemical enhanced oil recovery (EOR) has been adjudged as an efficient oil recovery technique to recover bypassed oil and residual oil trapped in the reservoir. This EOR method relies on the injection of chemicals to boost oil recovery. Recently, due to the limitations of the application of chemical EOR methods to reservoirs having elevated temperatures and high salinity and hardness concentrations, nanotechnology have been applied to enhance its efficiency and improve oil productivity. The synergistic combination of nanoparticles and conventional EOR chemicals has opened new routes for the synthesis and application of novel materials with sterling and fascinating properties. In this chapter, an up-to-date synopsis of nanotechnology applications in chemical EOR is discussed. A detailed explanation of the mechanism and applications of these novel methods for oil recovery are appraised and analyzed. Finally, experimental and laboratory results were outlined. This overview presents extensive information about new frontiers in chemical EOR applications for sustainable energy production.

Keywords: chemical enhanced oil recovery, nanotechnology, surfactant, polymer, nanoparticles

1. Introduction

Crude oil has remained the major source of world energy supply despite considerable efforts on other sources of energy [1]. Due to rapid industrialization, there is an increase in world energy demand leading the need to produce increasing volume of crude oil to support this demand. Meanwhile, the oil and gas industry is concerned with the shortage of new conventional oil reserves and low production from existing conventional reservoirs. On average, one-third of conventional reservoirs can be recovered through primary and secondary (i.e. waterflooding) oil recovery processes. The remaining oil-in-place is the target for enhanced oil recovery (EOR). Several EOR methods have been developed to recover bypassed and residual oil in the reservoir. These are majorly categorized into thermal and non-thermal EOR methods. Thermal EOR are unsuitable for reservoirs with great depth or thin pay zones. Hence, non-thermal EOR methods such as gas flooding, chemical flooding and microbial methods have received important attention over the last decades for oil recovery processes [2–4].

Of the numerous EOR methods, chemical EOR has been considered as the most promising because of its high efficiency, technical and economic feasibilities. Chemical EOR methods increase the efficiency of oil production by increasing the volumetric sweep efficiency of the injected waterflood. By tuning the efficiency of the injected chemical floods, the microscopic (pore scale) displacement efficiency and/or macroscopic (sweep) efficiency of the reservoir is increased leading to an increase in oil production. Chemicals for injection include alkali, surfactants, and polymers. Alkali and surfactants increase oil recovery by improving microscopic displacement at the pore scale; while polymers enhance the volumetric sweep efficiency of the reservoir [5].

Despite its highly reported efficiency and widely acclaimed potentials, chemical EOR has several limitations. The chemicals injected degrade and/or precipitate in the presence of resident reservoir brines and elevated temperature conditions. Besides, retention of the chemicals occurs during their flow in porous media which decreases it process efficiency and may lead to formation damage. To overcome this shortcoming, new salt and temperature-tolerant chemicals of various kinds have been developed and tested for their EOR potentials. Nonetheless, most of the newly developed chemicals have been jettisoned as they were found to increase the cost of the overall EOR process.

Nanotechnology is the application of nanoparticles characterized by a size ranging from 1 to 100 nm (see **Figure 1**) [6, 7]. In the oil and gas industry, applications of nanotechnology ranges from drilling processes, flow assurance, hydraulic fracturing, cementing, to EOR [8]. For EOR process, the engineered nanomaterials are mixed with fluids that are injected into the reservoir to boost oil production [9].

Nanoparticles and conventional EOR chemicals blends have been reported to possess important properties that are not observed in the individual chemical or nanoparticle [10]. For example, surfactant nanofluids (or nanosurfactant), a blend of nanoparticle and surfactant were reported to improve the efficiency of the surfactant at lowering the interfacial tension (IFT) of oil/water (o/w) interface and lower their adsorption during their transport in porous media [11]. Besides, emulsions and foams stabilized by nanoparticles are found to be thermodynamically

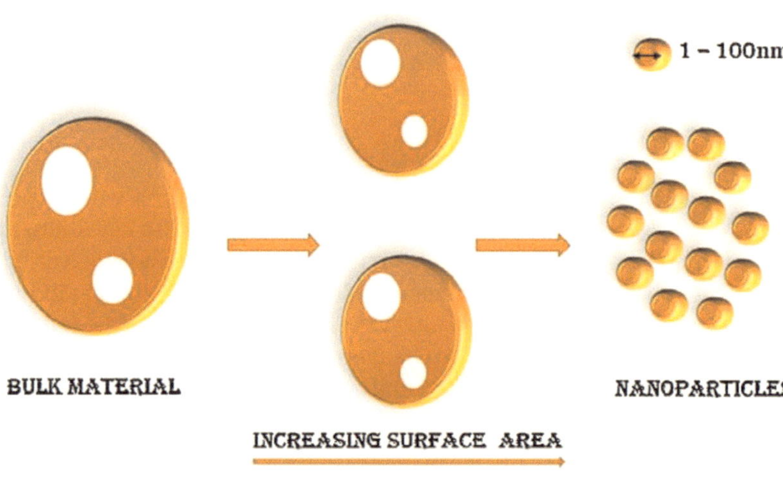

Figure 1.
Schematic of increasing surface area of nanoparticle with decreasing particle size [6].

stable and easily transported in reservoirs [12]. Meanwhile, polymeric nanofluids demonstrated to have improved rheological behavior and stability at characteristic reservoir temperature and salinity conditions [4]. This chapter presents an overview of nanotechnology applications in chemical EOR. First, the challenges of chemical EOR are briefly discussed. Subsequently, the mechanism and efficiency of nanotechnology application in chemical EOR is discussed. Finally, the experimental and laboratory studies of the newly devised EOR technique are outlined.

2. Challenges of chemical EOR methods

2.1 Degradation and precipitation

An oil reservoir exists at a specific temperature, salinity, and pH. The prevailing conditions of the reservoir influence the efficiency of the injected chemicals and consequently of the EOR process [13]. Most injected chemicals degrade and become unstable at high salinity, elevated temperature, and low pH conditions [14]. For polymers, under saline conditions, screening of the charged polymer molecules by cations contained in the reservoir brine occurs. This reduces the hydrodynamic radius and polymer chain entanglement causing the contraction of the macromolecules that ultimately results in the loss of polymer solution viscosity [15, 16]. Meanwhile, high temperature causes hydrolysis of the polymer and its precipitation in the presence of divalent ions [17]. In the case of surfactant and alkali solutions, depending on the rock type, precipitation of the chemicals occurs in the presence of divalent cations [18]. Low pH reservoir conditions might interact and acidify injected chemical solutions [15].

2.2 Adsorption and retention

Depending on the type of chemical injected into the reservoir, adsorption and retention of chemical occur during flow through porous media, which negatively affects the efficiency of the EOR process [19]. Chemicals react with the rock surface through electrostatic attraction, steric interaction, and van der Waal forces that

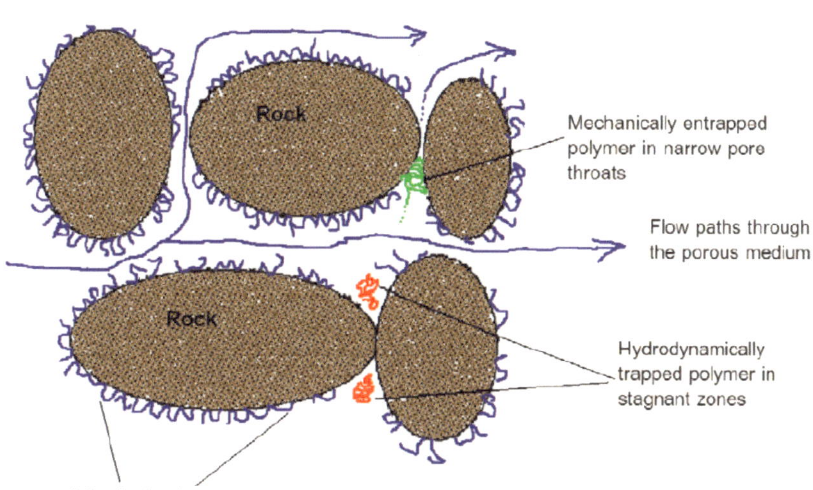

Figure 2.
Retention of polymers in porous media. Sourced from [19].

reduces the concentration of the injected chemical solutions. Adsorption is prevalent for surfactant and alkali chemicals, while polymer is mainly retained due to mechanical entrapment because of the size of the polymer macromolecules [3, 18, 20]. The adsorption process occurs when the interface is energetically favored by the surfactant and/or alkali in comparison to the bulk phase. Thus, the adsorption at the solid–liquid interface takes place by the transfer of the molecule of the chemical to the solid–liquid interface from the bulk solution phase [21]. Meanwhile, polymer retention and inaccessible pore volume dictates the propagation of polymer flow in the reservoir [22]. Retention of polymer is alluded to any mechanism that leads to reduction or removal of polymer molecules from transported aqueous phase. The nature of polymer retention in reservoir rock is depicted in **Figure 2**. Overall, adsorption and/or retention of chemicals in porous media governs the efficiency and economic viability of the EOR process. Several factors affecting adsorption or retention of chemical EOR includes; electrolyte concentration (salinity), temperature, pH, composition of reservoir fluids, and the presence of clay mineral content [21, 23].

3. Application of Nanotechnology in chemical EOR

Nanotechnology application in chemical EOR is used to overcome the shortcomings and improve the process efficiency of chemical EOR methods. Though most works are still at the laboratory scale, the synergic application of nanoparticles and chemicals have led to the formation of novel nanomaterials with exceptional qualities [6]. Recently, field trials have been reported in Columbia oilfield [24]. Depending on the nanoparticle type and chemical used, the formulated nanomaterials have demonstrated better stability and superior quality which enhances their performance during simulated reservoir conditions. So far, the most common nano-chemical studies are polymeric nanofluids and surfactant nanofluids.

3.1 Surfactant nanofluid

Surfactant nanofluid, a combination of nanoparticle and surfactant, increases the microscopic displacement efficiency through the mechanisms of IFT reduction and wettability alteration [11, 25, 26]. This nanofluid could be used for the generation or formation of stable foams and emulsions in the reservoir. Stable foams ensure fluid diversion from thief zones to lower permeability regions in the reservoir, while emulsions ensure conformance efficiency of the injectant [27, 28]. Furthermore, surfactant nanofluids have been reported to have lower adsorption onto rock surface compared to ordinary surfactant solutions [29, 30].

IFT and wettability are major parameters for quantifying fluids distribution and movement in the reservoir [31]. After secondary oil recovery, a portion of the oil is trapped in the reservoir due to capillary forces. This capillary force is measured by a dimensionless capillary number defined as [32]:

$$N_c = \frac{\mu v}{\sigma . \cos \theta} \tag{1}$$

Where μ is the displacing fluid viscosity, v is the displacing Darcy velocity, θ is the contact angle, and σ is the IFT between the displacing fluid (water) and the displaced fluid (oil). N_c is closely related to residual oil saturation and oil recovery. N_C increases as residual oil saturation decreases. Consequently, a higher N_c will result in a higher oil recovery [32, 33]. The capillary trapped oil can be released by either lowering the IFT of the o/w interface or through the wettability alteration mechanism [33]. Surfactant molecules due to their amphiphilic nature lowers IFT and alters wettability of porous media by adsorbing at the o/w interface [5, 32]. Nanoparticles also performs the same task in similitude to surfactant molecules, though at a different efficiency. The synergic effect of blending nanoparticles and surfactants have been shown to enhance the surfactant flooding process by lowering IFT and altering the wettability more efficiently than the individual nanoparticle or surfactant solution. On the other hand, the surfactant enhances the stability of the nanoparticles, thus, increasing the efficiency [11].

Driven primarily by electrostatic interaction, the surfactant adsorbs on the nanoparticles surface forming surfactant-coated nanoparticles [11]. Nonetheless, the relative concentration of nanoparticles and surfactant in the solution determines the properties of the mixture. A lower concentration ratio of surfactant to nanoparticle in the mixture means that only a small fraction of the nanoparticle surface will be coated by surfactant. Conversely, a higher concentration ratio of surfactant to nanoparticles implies the surfactant molecules will form a bilayer on the particle surface [11, 34]. A single-chain surfactant on nanoparticle is required to form maximum nanoparticle flocculation and hydrophobic nature required for an optimal performance.

3.1.1 IFT reduction

To quantify the performance of surfactant nanofluid on IFT of o/w interface, Le et al. evaluated the impact of silica (SiO_2) nanoparticles and anionic surfactant for IFT reduction using a spinning drop tensiometer. Their results indicated that at a total concentration of 1000 ppm and at a surfactant to SiO_2 ratio of 8:2, a four-fold IFT reduction was achieved by the nanosurfactant. Hence, they proposed the application of surfactant nanofluids for EOR in high temperature and high salinity (HTHS) conditions [35]. Mohajeri et al. evaluated the effect of zirconium oxide (ZrO_2) nanoparticles on anionic sodium dodecyl sulfate (SDS) surfactant and cationic cetyltrimethylammonium bromide (CTAB) surfactant [25]. They reported that the contribution of ZrO_2/SDS yielded an IFT reduction of 81% while ZrO_2/CTAB decreased the IFT of o/w interface by 70%. Zargartalebi et al. probed the effect of SiO_2 nanoparticles and anionic SDS to quantify the effect of the nanoparticle on IFT, adsorption and oil recovery potential of the surfactant molecules. They observed that nanoparticles effectively improve surfactant performance by enhancing the governing mechanism. Furthermore, flooding results shows that oil recovery increased significantly due to the inclusion of nanoparticles in the surfactant solution [26].

The mechanisms of nanosurfactant for reducing IFT at o/w interface has been explored. Researchers noted that the adsorption of the surfactant on the nanoparticle surface occurs as a result of the mix, leading to a hydrophobic character of the nanoparticle surface. Due to their Brownian motion, the nanoparticle acts as carriers for the surfactant molecules from the bulk of the fluid to the interface. At the interface, the minimization of the interfacial energy by the nanosurfactant leads to IFT reduction. As compared to ordinary surfactant whose molecule desorbs from the interface after some time, the nanoparticle prevents desorption of surfactant molecules from the interface, hence, better IFT reduction [11].

3.1.2 Wettability alteration

The reduction of interfacial energy at the rock/oil/brine interface by nanosurfactant also results in higher wettability alteration. Besides, the relative permeability curves of oil and water also changes after contact with the surfactant nanofluid; the relative permeability to water and oil decreases and increases respectively [34]. Mohajeri et al. studied the effect of ZrO_2/surfactants on wettability alteration in a fractured micromodel. The sessile drop experiments and wettability alteration measurements showed that coating the micromodel with heavy oil makes an oil-wet surface. Moreover, coating of the oil-wet micromodel with surfactant or nanoparticle altered the wettability of the surface to water-wet condition, while coating the surface with the blend ZrO_2/surfactant altered the wettability to strongly water-wet condition [25].

Additionally, the use of nanosurfactants as wettability alteration agents have proved useful for improving oil recovery from carbonates reservoir, which are characterized by poor oil recovery owing to its inherent natural fractures and hydrophobic nature that makes water imbibition into its rock matrix difficult, because of capillary pressure effects. Nwidee et al. assessed the effect of nanosurfactant formulation for wettability alteration of oil-wet limestone over a wide range of temperatures (0–70°C). ZrO_2 and nickel oxide (NiO) were used as the nanoparticles while CTAB and triton X-100 were used as surfactants for the formulation. Due to their strong electrostatic interaction, ZrO_2/CTAB and NiO/CTAB display greater affinity for the rock surface and demonstrated better wettability alteration efficiency for all temperature conditions considered in this study [36].

Surfactant nanofluids have also been used to improve wettability alteration of sandstone cores to boost oil recovery. Giraldo et al. tested alumina-surfactant nanofluid to improve oil recovery in sandstone cores via wettability alteration using contact angle and imbibition tests. Their results show that the effectiveness of surfactant as wettability modifier was improved with the addition of 100 ppm of alumina nanoparticles. Additionally, the effective oil permeability increased by 33%, and consequently, a higher oil recovery was recorded [37]. Huibers et al. measured changes in wettability of sandstone cores of saturated with light and heavy crude oil using surfactant nanofluid composed of SiO_2 nanoparticle and Tween 20 nonionic surfactant in Berea and Boise sandstone cores. Using direct imaging and contact angle measurements, 0.001 wt.% SiO_2 nanoparticles yielded an increase in contact angle of 101.6% for light oil saturated cores, while the optimum concentration for heavy oil was not ascertained at the nanoparticle concentration range investigated [38].

3.1.3 Adsorption reduction

One of the major challenges of surfactant EOR is the loss of surfactant molecules to adsorption onto the formation rock during the flooding process. Surfactant adsorption can make the chemical EOR process economically unfeasible. Therefore, reducing the surfactant adsorption improves the oil recovery process. Previous studies have investigated the use of polymers such as sodium polyacrylate as sacrificial agent to reduce surfactant adsorption during flow in porous media [39]. Recently, the adsorption reduction effect of nanoparticles has been investigated during the co-injection of nanoparticles with surfactant for oil recovery. Nanoparticles showed good potential for inhibiting surfactant adsorption via competitive adsorption mechanism by blocking the active site of the porous media while the surfactant flows through the porous media contacting the resident fluids in the reservoir. Yekeen et al. observed that the presence of SiO_2 and Al_2O_3 nanoparticles decreased surfactant adsorption on kaolinite in the presence of reservoir brines. The addition of Al_2O_3 nanoparticles reduced the SDS adsorption on kaolinite by 38%, while the addition of SiO_2 nanoparticles reduced the SDS adsorption by 75% [40].

Wu et al. conducted static and dynamic adsorption experiments to investigate the inhibition mechanism of SiO_2 nanoparticle during co-injection with surfactant. An optimum aging time, solid–liquid ratio, nanoparticle concentration, and surfactant concentration were determined for the adsorption process. Static adsorption experiments showed that 0.5 wt.% of SiO_2 nanoparticle concentration reduced the adsorption of SDS from 2.84 to 1.61 mg/g. Dynamic adsorption experiments conducted at 20°C showed that the adsorption and dynamic retention of single SDS solution were 1.16 and 0.30 mg/g respectively. The addition of 0.3 wt.% SiO_2 nanoparticle concentration reduced the adsorption and dynamic retention of the surfactant by 43.6% to 0.66 and 0.06 mg/g, respectively. Furthermore, oil recovery displacement experiments in sand packs using a nanosurfactant solution showed a 4.68% incremental oil recovery factor over the oil recovery attained by the injection of conventional surfactant solution [29]. Suresh et al. extended the adsorption studies on nanosurfactant to higher salinity and elevated temperature (80°C) conditions using thermogravimetric analysis (TGA) to determine the surfactant concentration in the produced fluids. The dynamic adsorption of the surfactant was calculated from the difference between injected and effluent concentration of the surfactant. Results showed that the addition of 0.5 wt.% SiO_2 nanoparticle to the surfactant solution reduced the surfactant adsorption by a factor of 3 times from 0.810 to 0.265 mg/g [30].

3.1.4 Foam and emulsion stability

Foams used for oil recovery are generated by co-injecting a gas (e.g., carbon dioxide, nitrogen or air) and a foaming agent containing liquid into the reservoir [41]. In the porous media, foams act as a dispersion of gas in liquid separated by a lamella, with the gas phase residing in the upper side while the bulk liquid is located at the bottom of the foam structure [42]. They perform two diverse roles in the reservoir namely; (1) mobility control, (2) fluid diversion. These mechanisms aid foam to overcome the challenges of gas EOR such as gravity override and viscous fingering phenomena. The liquids used as conventional foaming agents includes surfactants, polymers, and proteins. When used with polymers, foams are used to plug high permeability areas, while the polymer is diverted to lower permeability regions, thus, improving the volumetric sweep efficiency of the reservoir. In the case of surfactant-stabilized foams, a stable foam is formed due to a decrease in the required energy to form the gas–liquid interface [43]. Moreover, the synergic combination with surfactant lowers the interfacial tension of the capillary trapped oil, hence, facilitates oil displacement [44, 45]. Conventional foams have been shown to be thermodynamically unstable in the presence of oil and resident reservoir brines. This implies that the foam coalesces leading to of the reduced efficacy of the process. The addition of nanoparticles to the surfactant solution seems to generate more stable foams with longer half-life and ability to withstand harsh reservoir conditions [43]. Due to the solid nature of nanoparticles, the foams they stabilize are highly resistant to unfavorable reservoir conditions. Nanoparticles adsorb at the lamellae interface of the foam with a strong adhesion energy that makes their attachment irreversible (see **Figure 3**) [41, 46].

Sun et al. studied the influence of nanoparticles on the generation, propagation, and stability of SiO_2/SDS-stabilized foam in micromodels and sandpack porous media [47]. In the case of the SDS-stabilized foam, the shape of the oil droplet could not be changed by the foam because the microforce acting on the oil droplet was small. This subsequently leads to bubbles rupture and coalescence leaving a substantial amount of oil trapped in the porous media. In the case of SiO_2/SDS foam, a

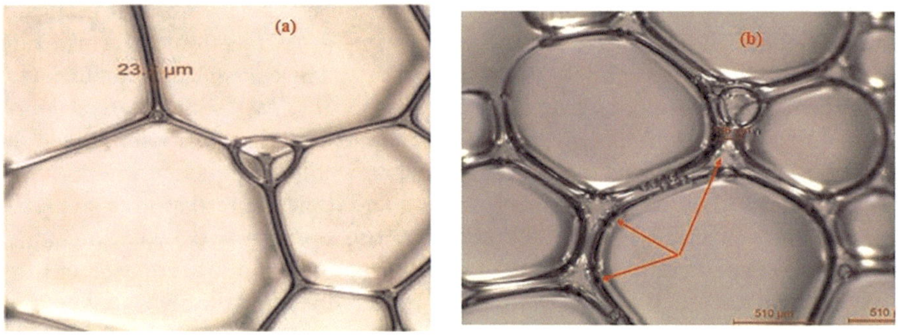

Figure 3.
Foams stabilized (a) without nanoparticles showing signs of foam drainage, (b) with nanoparticles stabilizing the lamellae [46].

large amount of oil was displaced by the foam due to the higher microforce acting on the oil droplet. The higher microforce was attributed to the enhanced viscoelasticity of the bubble surface by the attached nanoparticles. Yekeen et al. studied the influence of SiO_2 and Al_2O_3 nanoparticles on surfactant-foam stability and propagation in the presence of oil. They noted that the presence of nanoparticle increases foam half-life. Additionally, the SiO_2-SDS and Al_2O_3-SDS foam achieved nearly 100% microscopic efficiency even in the presence of oil. Finally, they identified mechanisms of foam flow as lamellae division and bubble-to-multiple bubble lamellae division, while the dominant mechanism of oil displacement and residual oil saturation are direct displacement and emulsification of oil [40]. **Tables 1** and 2 summarizes laboratory and experimental results of nanoparticles-stabilized and nanoparticle-surfactant stabilized foams.

On the other hand, the interactions of surfactants with oil during flow through porous media may generate emulsions. Emulsions generated *in situ* have potential for mobility and conformance control in the reservoir, which are desirable properties for improving the oil recovery process. Further, the feasibility of injecting emulsions has also been explored, exhibiting appropriate potential for oil recovery. Nonetheless, conventional emulsions show poor stability at high pressure and high temperature conditions. As temperature increases, the average droplet size of the dispersed phase increases which eventually plug pore throats in the reservoir [75]. Recently, the binary mix of nanosurfactant with oil have been evaluated for emulsion generation showing better stability performance in the reservoir for oil recovery applications. Besides, the presence of nanoparticle significantly improved the stability and mobility of the emulsions.

Pei et al. investigated the synergetic effect of SiO_2 nanoparticle and CTAB for o/w emulsions applications. Phase behavior testing, rheology evaluation, and microvisualization studies showed that nano-surfactant-stabilized emulsion demonstrated a high bulk viscosity and desirable mobility for recovering heavy oil [50]. Kumar et al. synthesized a Pickering emulsion stabilized by SiO_2 nanoparticle and sodium dodecylbenzene sulfonate (SDBS) surfactant. The synthesized Pickering emulsion displayed better thermal stability at the high pressure (0–5 MPa) and high temperature (30–100°C) conditions investigated, and showed a higher oil recovery when injected into the sandpack [75].

3.2 Polymeric nanofluids

Depending on the method of preparation, polymeric nanofluids are categorized into two types; polymer-coated nanoparticles and polymer nanoparticles.

Nanoparticle	Surfactant	NP conc. (wt.%)	Surfactant conc.	Base fluid	Oil type	Porous media type	Mechanism of recovery	RF	Ref.
SiO_2	SDS	0.1–0.5	0.2 wt.%	Deionized water	Crude oil	Quartz sand	IFT, competitive adsorption	4.68% IO	[29]
SiO_2	Alfoterra, Soloterra	0.1–0.8	0.005–0.2 wt.%	Brine	Crude oil	Dolomite, Limestone	IFT, WA	37–45% OOIP	[48]
SiO_2	TX-100	0.1	0.1 wt.%	Brine	Crude oil + kerosene	Sandstone	IFT, WA	8% IO	[49]
SiO_2	CTAB	0.05–0.5	0.1 wt.%	Brine	Heavy oil	Micromodel	ES	17.4–38% OOIP	[50]
SiO_2	SDS	0.25	1 wt.%	Brine	Heavy oil	Mircomodel	IFT, ES	5% IO	[51]
SiO_2	Non-ionic Tween 20	2–4	0.74 wt.%	Deionized water	Heavy oil	Micromodel	IFT, ES	18–39%	[52]
SiO_2	CTAB	0–5	0.09 mM	Distilled water	Heptane	—	IFT	—	[53]
Hydrophilic and hydrophobic SiO_2	SDS	0.1	0.01–0.6 wt.%	Deionized water	—	Sandstone	IFT, Adsorption reduction	—	[54]
Non-ferrous metal	Sulphanole	0.001	0.0078–0.05 wt.%	—	Heavy oil	—	IFT, WA	12–22% IO	[55]
Al_2O_3	PRNS	0.001–1		Distilled water	Heavy oil	Sandstone	WA	33% IO	[37]
Fumed SiO_2 Hydrophobic SiO_2	Zyziphus Spina Christi	0.05–0.2	0.1–8 wt.%	Deionized water	—	Shale sandstone	Adsorption reduction	—	[56]
ZrO_2	SDS, CTAB	0.01	0.1–0.4 wt.%	Distilled water	Heavy oil	Micromodel	IFT, WA	—	[25]
ZrO_2, NiO	TX-100, CTAB	0.004–0.05	0.1–3.2 wt.%	Deionized water	—	Limestone	WA	—	[36]
SiO_2	SDS, PAM	0.5–2.0	0.14 wt.%	Deionized water	Medium oil	Sandstone	IFT, WA	60% OOIP	[57]

Table 1.
Laboratory results of oil recovery applications by nanosurfactant.

Ref.	NP type	NP size (nm)	NP Conc.	Surfactant	Temperature	Pressure	Salinity	Foam generator	Oil recovery
[58]	SiO_2, Al_2O_3, CuO, TiO_2	10–40	0.002–0.1 wt.%	—	27°*C*	—	NaCl (0.3 wt%)	Sandpack	5.1–17.4% OOIP[a]
[59]	PEG coated-SiO_2	5	0.01–0.1 wt.%	—	21.1–90°*C*	1350–1400 psia	NaCl (2–4 wt.%)	Glass beads pack, Capillary tube	—
[60]	SiO_2	100–150	0 – 5 wt.%	—	25°*C*, 60°*C*	1200 – 2000psia	NaCl (0.5, 2.0, 5.0%)	Sapphire observation tube	—
[61]	AlOOH	10–100	1 wt.%	SC (0–100 mM)	60°*C*	6 mPa	NaCl (10–600 mM), $CaCl_2$ (10–200 mM)	Sanpack	~20% OOIP[a]
[62]	SiO_2	17	0.01–0.5 wt.%	—	25°*C*	1200 psig, 1500 psig	NaCl (2.0%)	Berea sandstone core	—
[63]	SiO_2	100–200	0.05–3.0% w/v	PEG, Tergitol 15-S-20, DCDMS	35°*C*, 50°*C*	1200–3000 psia	—	Glassbead pack	—
[64]	APTES – SiO_2	20–30	0.01 wt.%	SDS (0.4 wt.%)	25°*C*	14.7 psi	—	Glassbead pack	18% OOIP[a]
[65]	PECNP	—	1.0 wt.%	Surfonic N120	40°*C*	1300 psi, 1800 psi	KCl (2.0 wt.%)	Indiana limestone	10.71% OOIP[b]
[66]	TTFA	80	0.5 wt.%	Cationic, anionic and non-ionic surfactant (0.2 wt.%)	25°*C*	1300 psi (backpressure)	NaCl (1.0 wt.%)	Berea sandstone	—
[67]	PEG coatedSiO_2	10 (20)	0.5 wt.%	AOS (0–0.5 wt.%)	25°*C*	100 psi (backpressure)	NaCl (1–10 wt.%), API Brine	Berea sandstone	10% OOIP[a]
[68]	SiO_2, Al_2O_3, CuO, TiO_2	10–40	0.1–1.0 wt.%	AOS (O.5 wt.%)	Room temperature	—	NaCl (2 wt.%)	Sandpack models	5–14% OOIP[a]
[69]	FA	100–200	< 0.05 wt.%	AOS (0.0315 wt.%)	—	—	NaCl (1.0–5.0 wt.%), $CaCl_2$ (0–9.5 wt.%)	Bentheimer sandstone	—
[70]	SiO_2	5, 12, 25, 80	0.5–10 wt.%	—	70°*C*	2200 psi	NaCl (8.0 wt.%), $CaCl_2$ (2.0 wt.%)	Biose sandstone, Sandpack	—

Ref.	NP type	NP size (nm)	NP Conc.	Surfactant	Temperature	Pressure	Salinity	Foam generator	Oil recovery
[71]	Al_2O_3 – coated SiO_2	20	1–5 wt.%	Triton CG-110 AOS, PG (0.1–0.5 wt.%	Room temperature	100 psi (backpressure)	—	Berea sandstone	14.8–20.6% OOIP[a]
[72]	MWCNT	10	0.01 wt.%	Tergitol 15-s-40, AOS	25°C	—	NaCl (2.4 wt.%), $CaCl_2$ (0.6 wt.%)	Ottawa sandpack	—
[73]	PEG coated SiO_2	5 (10)	0.3 wt.%	AOS (0.5 wt.%	55, 75°C	110 psi (backpressure)	NaCl (1–8 wt.%)	Heterogeneous sandpack	34.4% OOIP[a] 9% OOIP[b]
[46]	SiO_2 Al_2O_3	12–20	0.05–5.0 wt%	SDS (0.03 wt.%)	25°C	—	NaCl (0.5 wt.%)	Hele-shaw cell	—
[74]	PEG-SiO_2, GLYMO-SiO_2	12, 20	0.5 wt.%	AOS	25, 60, 80°C	110 psi (backpressure)	NaCl (8 wt.%), $CaCl_2$ (2 wt.%)	Sandpack	29.0–43.3% OOIP[a]

[a]*Incremental oil recovery over waterflood.*
[b]*Incremental oil recovery over surfactant-stabilized foam.*

Table 2.
Laboratory and experimental investigation of nanoparticle/nanoparticle-surfactant stabilized foams.

Polymer-coated nanoparticles were developed due to overcome the aggregation and agglomeration problems of nanoparticles at reservoir conditions. It involves grafting polymers onto the surface of nanoparticles to improve dispersibility. In addition, their properties can be customized for particular applications [34]. Meanwhile, polymer nanoparticles are prepared by the hybrid dispersion of nanoparticles in polymer solutions. These polymer nanoparticles emerged as a means of inhibiting polymer degradation in typical reservoir conditions [4]. The mechanisms of polymeric nanofluids performance during EOR applications include improved rheology and stability, wettability alteration, and lower polymer adsorption [6].

3.2.1 Improved rheology and stability for mobility control

Rheology is defined as the study of flow and deformation behaviors of fluids under stress [76]. For EOR applications, an improved rheological behavior of injectant is required to inhibit viscous fingering phenomena and maintain a suitable mobility ratio in the reservoir; which requires that the displacing fluid maintain its viscosity and chemical integrity in the presence of resident reservoir brines [77]. Polymer and nanoparticles undergo degradation in the presence of reservoir brines. The cations present in the brine interact with the carboxylate and amide groups in the polymer molecule resulting in viscosity loss [78]. In the case of nanoparticles, the electrostatic attraction among nanoparticles are increased in the presence of brine fostering their aggregation and agglomeration; which implies the loss of surface functionality that is required for EOR [79]. However, the combination of polymer and nanoparticles results in a synergistic effect that improves the rheology of the polymer and the stability of the nanoparticle [4].

The preparation of polymer-nanoparticles blends involves the mixing of the nanoparticle and the polymer solution or grafting of the polymer on the nanoparticle [80, 81]. Subsequently, interactions occur between the nanoparticle and the carboxylate and amide group in the polymer molecules. Therefore, nanoparticles act as physical crosslinkers among the polymer chains forming three-dimensional network of stable flocs that increases the viscosity of the suspension [82]. At high temperature, polymer-nanoparticles blends exhibit better rheological performance due to the enhanced bridging induced flocculation [82, 83]. Furthermore, in the presence of reservoir brines, nanoparticles shield the polymer backbone from the cations of the brine by inducing ion-dipole interactions that inhibit the degradation of the polymer molecules [81].

Lai et al. noted that the shear and mechanical resistance of acrylamide polymer solution can be increased by adding modified nano-SiO_2, because the presence of SiO_2-NP caused a reduction of the hydrodynamic radius of the polymer molecules [84]. Hu et al. studied the rheological properties on an oilfield polyacrylamide (HPAM) -SiO_2 NP under different aging times, salinity, and temperature conditions. The results demonstrated that the presence of the SiO_2-NP significantly improved the viscosity and viscoelastic properties of the HPAM under high temperature and high salinity (HTHS) conditions [80]. Haruna et al. grafted HPAM molecules with graphene oxide (GO) nanosheets and evaluated the rheological and stability properties of the formulated polymeric nanofluid. They reported enhancement of the suspension viscosity behavior, as well as high-temperature stability and improved elastic properties of the dispersion [85].

As for polymer-coated nanoparticles, depending on the grafting method, the polymeric chains protrude from the nanoparticle surface. Hence, hydrodynamic interactions occur between the grafted nanoparticle when subjected to shear. Besides, polymeric chains grafted on the surface of the nanoparticle overlap with another polymer chain adsorbed on another nanoparticle. The overlapping of several grafted nanoparticles results in the strengthening of the network structure of the polymer -nanoparticle system. Consequently, hydro clusters are formed, which results in an increase of stability and viscosity [86]. Liu et al. grafted a layer of amphiphilic-polymeric chains on nano-SiO_2 core shell via a facile water-free radical polymerization and evaluated its rheological properties and oil recovery performance. The synthesized polymer-coated nanoparticle formed a three-dimensional microstructure and intermolecular associations characterized by long-term stability and better rheological properties than the individual polymer or nanoparticles. Furthermore, a 20% incremental oil recovery was recorded after flooding the polymer-coated nanoparticle solution at a concentration of 1500 mg/L in sandstone cores [87]. **Table 3** summarizes some laboratory and experimental studies of improved rheological properties and oil displacement properties of polymeric nanofluids.

3.2.2 Adsorption inhibition

Polymeric nanofluids also show reduced adsorption onto porous media due to the synergic interaction between the polymer and nanoparticles. Foster et al. used the grafting through approach to tether tuneable quantities of poly(2-acrylamido-2-methylpropanesulfonic acid) (PAMPS) and poly([3-(methacryloxylamino)propyl] dimethyl(3-sulfopropyl)ammoniumhydroxide)(PMPDSA) homopolymer (PMPDSA) onto iron oxide nanoparticle surfaces. Steric stabilization of the synthesized polymer-coated nanoparticle was observed which remained stable at HTHS conditions. Moreover, adsorption experiments on crushed Berea sandstone cores showed that the adsorption of polymer-coated iron oxide nanoparticles was

References	NP type	Polymer/ copolymer type	PNF conc. (ppm)	Brine/conc.	Temp. (°C)	Porous media type	Incremental oil recovery (%)
[88]	SiO_2	PEOMA	10,000	1.0 wt.% NaCl	30	Berea sandstone	19.5
[89]	SiO_2	AMPS	50,000	—	80	Quartz sand	23.22
[90]	SiO_2	PEG	10,000	—	80	Glass micromodel	20.0
[87]	SiO_2	$MeDiC_8AM$	1500	12 wt.% (NaCl & $CaCl_2$)	82.3	Sandstone	20.0
[91]	SiO_2	$AMC_{12}S$	1100	18 wt.%	110	Sandstone	24.0
[92]	SiO_2	AA/AM	2000	2 wt.% NaCl, 0.18 wt.% $CaCl_2$	65	Sandstone	20.1
[93]	SiO_2	AM/AA	1500	—	—	—	18.84
[94]	SiO_2	HPAM	1000	2.4 wt.% (NaCl, $CaCl_2$, $MgCl_2$)	25	Glass micromodel	10.0
[95]	SiO_2	HPAM	800	3 wt.% NaCl	—	Glass micromodel	10.0
[76]	MMT clay	HPAM	1000	10 wt.% (NaCl, $CaCl_2$, $MgCl_2$)	90	Quartz sand	33.0

Table 3.
Laboratory results of oil recovery applications by polymeric nanofluid [6].

infinitesimal and almost negligible [96]. Cheraghian et al. performed static adsorption experiments to investigate the impact of nano-SiO_2 and nanoclay on the adsorption inhibition of polyacrylamide onto sandstone rocks. Polymer nanoparticles containing SiO_2 nanoparticle showed lower adsorption onto sandstone rock surface compared to the polymer containing nanoclay [97].

3.2.3 Wettability alteration

Wettability alteration plays a vital role in enhancing the microscopic displacement efficiency. In the case of polymeric nanofluids, an interplay of electrostatic repulsive forces occur at the interface of the nanoparticles., Two-dimension layered structure of nanoparticles occur due to Brownian motion when brought into contact with an oil-wet solid surface, creating a wedge film because of the ordering of nanoparticles at the three-phase (solid-oil–water) contact region. This results in an increase of the disjoining pressure,which causes the spreading of the nanofluid phase at the wedge of the vertex, altering the wettability of the surface [6]. Maurya et al. grafted polyacrylamide on the surface of SiO_2 using the free radical polymerization approach and investigated its wettability potential on an oil-wet sandstone rock surface. They indicated that the polymer grafted nanoparticle altered the wettability of the sandstone surface to a more water-wet condition [86]. Maghzi et al. performed wettability alteration studies employing polymer nanoparticles consisting of SiO_2 nanoparticle and polyacrylamide polymer solution in a five-spot glass micromodel. The polymer nanoparticle altered the surface of the micromodel from an average contact angle of 112° (oil-wet) to 20° (water-wet). More details of wettability alteration by polymeric nanofluids can be found in the literature [6, 34].

4. Conclusions

This chapter summarizes some of the recent advances in the application of nanotechnology in chemical EOR processes to boost oil production. The mechanisms of oil recovery through nanotechnology were reviewed. Several experimental studies were summarized and discussed. Results of various experiments shows that the incorporation of nanotechnology with chemical EOR shows good potential to improve pore scale mechanisms in the case of surfactant. Adsorption of surfactant on rock pores is inhibited while greater IFT reduction and better wettability alteration were achieved. Furthermore, nanotechnology improved the rheological properties of polymer and stability of emulsions and foams indicating the good potentials of improving sweep efficiency of injected chemicals especially in the presence of harsh reservoir conditions. Finally, future research should focus on modeling the flow behavior of nanomaterials through porous media, which is required for the designing and field implementation of nano-chemicals EOR.

Nomenclature

AA	acrylic acid
AlOOH	aluminum hydroxide
AM	acrylamide
AMPS	2-acrylamido-2-methyl-1-propanesulfonic acid
$AMC_{12}S$	2-acrylamido-dodecylsulfonate
APTES	(3-aminopropyl)triethoxysilane

AOS	alpha olefin sulfonate
$CaCl_2$	calcium chloride
CuO	copper oxide
DCDMS	dichlorodimethylsilane
ES	emulsion stability
FA	fly ash
GLYMO	(3-glycidyloxypropyl)trimethoxysilane
HPAM	hydrolysed polyacrylamide
IO	incremental oil
KCl	potassium chloride
PAM	polyacrylamide
PECNP	polyelectrolyte composite nanoparticle
PEG	polyethylene glycol
PEOMA	poly(oligo(ethylene oxide) mono methyl ether methacrylate)
$MeDiC_8AM$	2-methyl-N,N-dioctyl-acrylamide
MWCNT	multiwall carbon nanotube
$MgCl_2$	magnesium chloride
MMT	montmorillonite
NaCl	sodium chloride
NiO	nickel oxide
NP	nanoparticle
OOIP	original-oil-in-place
RF	recovery factor
SC	sodium cumenesulfonate
TTFA	thermally treated fly ash
WA	wettability alteration

Author details

Afeez Gbadamosi[1,2]*, Radzuan Junin[1], Muhammad Manan[1], Augustine Agi[1] and Jeffrey Oseh[1,2]

1 Department of Chemical and Petroleum Engineering, School of Engineering, Afe Babalola University, Ado-Ekiti, Nigeria

2 Department of Petroleum Engineering, Universiti Teknologi Malaysia, Johor Bahru, Malaysia

*Address all correspondence to: gbadamosiafeezo@abuad.edu.ng

References

[1] Afolabi F, Ojo T, Udeagbara S, Gbadamosi A. Bitumen extraction from Tar Sands using solvent techniques. International Journal of Scientific and Engineering Research. 2017;**8**: 783-790

[2] Gbadamosi AO, Kiwalabye J, Junin R, Augustine A. A review of gas enhanced oil recovery schemes used in the North Sea. Journal of Petroleum Exploration and Production Technologies. 2018;**8** (4):1373-1387. DOI: 10.1007/s13202-018-0451-6

[3] Abbas AH, Sulaiman WRW, Jaafar MZ, Gbadamosi AO, Ebrahimi SS, Elrufai A. Numerical study for continuous surfactant flooding considering adsorption in heterogeneous reservoir. Journal of King Saud University—Engineering Sciences. 2018:1-9. DOI: 10.1016/j.jksues.2018.06.001

[4] Gbadamosi AO, Junin R, Manan MA, Yekeen N, Augustine A. Hybrid suspension of polymer and nanoparticles for enhanced oil recovery. Polymer Bulletin. 2019:1-38. DOI: 10.1007/s00289-019-02713-2

[5] Sheng JJ. A comprehensive review of alkaline–surfactant–polymer (ASP) flooding. Asia-Pacific Journal of Chemical Engineering. 2014;**9**:471-489. DOI: 10.1002/apj

[6] Gbadamosi AO, Junin R, Manan MA, Yekeen N, Agi A, Oseh JO. Recent advances and prospects in polymeric Nanofluids application for enhanced oil recovery. Journal of Industrial and Engineering Chemistry. 2018;**66**:1-16. DOI: 10.1016/j.jiec.2018.05.020

[7] Khan I, Saeed K, Khan I. Nanoparticles: Properties, applications and toxicities. Arabian Journal of Chemistry. 2017:1-24. DOI: 10.1016/j.arabjc.2017.05.011

[8] Ko S, Huh C. Use of nanoparticles for oil production applications. Journal of Petroleum Science and Engineering. 2019;**172**:97-114. DOI: 10.1016/j.petrol.2018.09.051

[9] Agi A, Junin R, Gbadamosi A. Mechanism governing nanoparticle flow behaviour in porous media: Insight for enhanced oil recovery applications. International Nano Letters. 2018;**8**:1-29. DOI: 10.1007/s40089-018-0237-3

[10] Ali JA, Kolo K, Manshad AK, Mohammadi AH. Recent advances in application of nanotechnology in chemical enhanced oil recovery: Effects of nanoparticles on wettability alteration, interfacial tension reduction, and flooding. Egyptian Journal of Petroleum. 2018;**27**:1371-1383. DOI: 10.1016/j.ejpe.2018.09.006

[11] Almahfood M, Bai B. The synergistic effects of nanoparticle-surfactant nanofluids in EOR applications. Journal of Petroleum Science and Engineering. 2018;**171**:196-210. DOI: 10.1016/j.petrol.2018.07.030

[12] Sun X, Zhang Y, Chen G, Gai Z. Application of nanoparticles in enhanced oil recovery: A critical review of recent Progress. Energies. 2017;**10**: 345. DOI: 10.3390/en10030345

[13] Mandal A. Chemical flood enhanced oil recovery: A review. International Journal of Oil, Gas and Coal Technology. 2015;**9**:241-264

[14] Gbadamosi AO, Junin R, Manan MA, Agi A, Yusuff AS. An overview of chemical enhanced oil recovery: Recent advances and prospects. International Nano Letters. 2019:1-32. DOI: 10.1007/s40089-019-0272-8

[15] Chul JJ, Ke Z, Hyun CB, Jin CH. Rheology and polymer flooding characteristics of partially hydrolyzed

polyacrylamide for enhanced heavy oil recovery. Journal of Applied Polymer Science. 2012;**127**:4833-4839. DOI: 10.1002/app.38070

[16] Thomas A. Polymer flooding. In: Romero-Zeron L, editor. Chem. Enhanc. Oil Recover.—A Pract. Overv. Rijeka: InTech; 2016. pp. 25-50. DOI: 10.5772/64623

[17] Kamal MS, Sultan AS, Al-Mubaiyedh UA, Hussein IA. Review on polymer flooding: Rheology, adsorption, stability, and field applications of various polymer systems. Polymer Reviews. 2015;**55**:491-530. DOI: 10.1080/15583724.2014.982821

[18] Olajire AA. Review of ASP EOR (alkaline surfactant polymer enhanced oil recovery) technology in the petroleum industry: Prospects and challenges. Energy. 2014;**77**:963-982. DOI: 10.1016/j.energy.2014.09.005

[19] Sorbie KS. Polymer-Improved Oil Recovery. Glasgow and London: Blackie and Son Ltd; 1991. DOI: 10.1007/978-94-011-3044-8

[20] Bessaies-Bey H, Fusier J, Harrisson S, Destarac M, Jouenne S, Passade-Boupat N, et al. Impact of polyacrylamide adsorption on flow through porous siliceous materials: State of the art, discussion and industrial concern. Journal of Colloid and Interface Science. 2018;**531**:693-704. DOI: 10.1016/j.jcis.2018.07.103

[21] Bera A, Kumar T, Ojha K, Mandal A. Adsorption of surfactants on sand surface in enhanced oil recovery: Isotherms, kinetics and thermodynamic studies. Applied Surface Science. 2013;**284**:87-99. DOI: 10.1016/j.apsusc.2013.07.029

[22] Idahosa PEG, Oluyemi GF, Oyeneyin MB, Prabhu R. Rate-dependent polymer adsorption in porous media. Journal of Petroleum Science and Engineering. 2016;**143**:65-71. DOI: 10.1016/j.petrol.2016.02.020

[23] Mishra S, Bera A, Mandal A, Mishra S, Bera A, Mandal A. Effect of polymer adsorption on permeability reduction in enhanced oil recovery. Journal of Petroleum Engineering. 2014;**2014**:1-9. DOI: 10.1155/2014/395857

[24] Franco CA, Zabala R, Cortés FB. Nanotechnology applied to the enhancement of oil and gas productivity and recovery of Colombian fields. Journal of Petroleum Science and Engineering. 2017;**157**:39-55. DOI: 10.1016/j.petrol.2017.07.004

[25] Mohajeri M, Hemmati M, Shekarabi AS. An experimental study on using a nanosurfactant in an EOR process of heavy oil in a fractured micromodel. Journal of Petroleum Science and Engineering. 2015;**126**:162-173. DOI: 10.1016/j.petrol.2014.11.012

[26] Zargartalebi M, Kharrat R, Barati N. Enhancement of surfactant flooding performance by the use of silica nanoparticles. Fuel. 2015;**143**:21-27. DOI: 10.1016/j.fuel.2014.11.040

[27] Rafati R, Haddad AS, Hamidi H. Experimental study on stability and rheological properties of aqueous foam in the presence of reservoir natural solid particles. Colloids and Surfaces A: Physicochemical and Engineering Aspects. 2016;**509**:19-31. DOI: 10.1016/j.colsurfa.2016.08.087

[28] Bera A, Mandal A. Microemulsions: A novel approach to enhanced oil recovery: A review. Journal of Petroleum Exploration and Production Technologies. 2015;**5**:255-268. DOI: 10.1007/s13202-014-0139-5

[29] Wu Y, Chen W, Dai C, Huang Y, Li H, Zhao M, et al. Reducing surfactant adsorption on rock by silica nanoparticles for enhanced oil recovery. Journal of Petroleum Science and

Engineering. 2017;**153**:283-287. DOI: 10.1016/j.petrol.2017.04.015

[30] Suresh R, Kuznetsov O, Agrawal D, Darugar Q, Khabashesku V. Reduction of surfactant adsorption in porous media using silica nanoparticles. OTC-28879-MS, Present. Offshore Technol. Conf. 30 April-3 May, Houston, Texas, USA. 2018. pp. 1-9

[31] Moghadam AM, Mahsa BS. Enhancing hydrocarbon productivity via wettability alteration: A review on the application of nanoparticles. Reviews in Chemical Engineering. 2018; **35**(4):531-563. DOI: 10.1515/revce-2017-0105

[32] Kamal MS, Hussein IA, Sultan AS. Review on surfactant flooding: Phase behavior, retention, IFT, and field applications. Energy & Fuels. 2017;**31**: 7701-7720. DOI: 10.1021/acs.energyfuels.7b00353

[33] Pal S, Mushtaq M, Banat F, Al Sumaiti AM. Review of surfactant-assisted chemical enhanced oil recovery for carbonate reservoirs: Challenges and future perspectives. Petroleum Science. 2018;**15**:77-102. DOI: 10.1007/s12182-017-0198-6

[34] ShamsiJazeyi H, Miller CA, Wong MS, Tour JM, Verduzco R. Polymer-coated nanoparticles for enhanced oil recovery. Journal of Applied Polymer Science. 2014;**131**:1-13. DOI: 10.1002/app.40576

[35] Le NYT, Pham DK, Le KH, Nguyen PT. Design and screening of synergistic blends of SiO_2 nanoparticles and surfactants for enhanced oil recovery in high-temperature reservoirs. Advances in Natural Sciences: Nanoscience and Nanotechnology. 2011;**2**:35013

[36] Nwidee LN, Lebedev M, Barifcani A, Sarmadivaleh M, Iglauer S. Wettability alteration of oil-wet limestone using surfactant-nanoparticle formulation. Journal of Colloid and Interface Science. 2017;**504**:334-345. DOI: 10.1016/j.jcis.2017.04.078

[37] Giraldo J, Benjumea P, Lopera S, Cortés FB, Ruiz MA. Wettability alteration of sandstone cores by alumina-based Nanofluids. Energy & Fuels. 2013;**27**:3659-3665. DOI: 10.1021/ef4002956

[38] Huibers BMJ, Pales AR, Bai L, Li C, Mu L, Ladner D, et al. Wettability alteration of sandstones by silica nanoparticle dispersions in light and heavy crude oil. Journal of Nanoparticle Research. 2017;**19**:1-18. DOI: 10.1007/s11051-017-4011-7

[39] ShamsiJazeyi H, Verduzco R, Hirasaki GJ. Reducing adsorption of anionic surfactant for enhanced oil recovery: Part I. competitive adsorption mechanism. Colloids and Surfaces A: Physicochemical and Engineering Aspects. 2014;**453**:162-167. DOI: 10.1016/j.colsurfa.2013.10.042

[40] Yekeen N, Manan MA, Idris AK, Samin AM, Risal AR. Experimental investigation of minimization in surfactant adsorption and improvement in surfactant-foam stability in presence of silicon dioxide and aluminum oxide nanoparticles. Journal of Petroleum Science and Engineering. 2017;**159**: 115-134. DOI: 10.1016/j.petrol.2017.09.021

[41] Yekeen N, Manan MA, Idris AK, Padmanabhan E, Junin R, Samin AM, et al. A comprehensive review of experimental studies of nanoparticles-stabilized foam for enhanced oil recovery. Journal of Petroleum Science and Engineering. 2018;**164**:43-74. DOI: 10.1016/j.petrol.2018.01.035

[42] Romero-Zeron LB, Kantzas A. The effect of wettability and pore geometry on foamed-gel-blockage performance. SPE Reservoir Evaluation and Engineering. 2007;**10**:150-163. DOI: 10.2118/89388-PA

[43] Babamahmoudi S, Riahi S. Application of nano particle for enhancement of foam stability in the presence of crude oil: Experimental investigation. Journal of Molecular Liquids. 2018;**264**:499-509. DOI: 10.1016/j.molliq.2018.04.093

[44] Osei-Bonsu K, Grassia P, Shokri N. Relationship between bulk foam stability, surfactant formulation and oil displacement efficiency in porous media. Fuel. 2017;**203**:403-410. DOI: 10.1016/j.fuel.2017.04.114

[45] Romero L, Kantzas A. The effect of wettability and pore geometry on foamed gel blockage performance in gas and water producing zones. SPE/DOE Symposium on Improved Oil Recovery. 2004;**17**:1-17. DOI: 10.2118/89388-MS

[46] Yekeen N, Idris AK, Manan MA, Samin AM, Risal AR, Kun TX. Bulk and bubble-scale experimental studies of influence of nanoparticles on foam stability. Chinese Journal of Chemical Engineering. 2017;**25**:347-357. DOI: 10.1016/J.CJCHE.2016.08.012

[47] Sun Q, Li Z, Li S, Jiang L, Wang J, Wang P. Utilization of surfactant-stabilized foam for enhanced oil recovery by adding nanoparticles. Energy & Fuels. 2014;**28**:2384-2394. DOI: 10.1021/ef402453b

[48] Haeri MF. Experimental Evaluation of Surfactant-based Nanofluids on Wettability Alteration and Oil Recovery. Louisiana State University Doctoral Dissertations; 2018;4756. Available from: https://digitalcommons.lsu.edu/gradschool_dissertations/4756

[49] Zhao M, Lv W, Li Y, Dai C, Wang X, Zhou H, et al. Study on the synergy between silica nanoparticles and surfactants for enhanced oil recovery during spontaneous imbibition. Journal of Molecular Liquids. 2018;**261**:373-378. DOI: 10.1016/j.molliq.2018.04.034

[50] Pei H, Zhang G, Ge J, Zhang J, Zhang Q. Investigation of synergy between nanoparticle and surfactant in stabilizing oil-in-water emulsions for improved heavy oil recovery. Colloids and Surfaces A: Physicochemical and Engineering Aspects. 2015;**484**:478-484. DOI: 10.1016/j.colsurfa.2015.08.025

[51] Mobaraki S, Zakavi M, Mahmoodi O, Omidvar Sorkhabadi M, Khalilinezhad SS, Shiri TR. An experimental study on the mechanisms of enhancing oil recovery by nanoparticles-assisted surfactant flood. Geosystem Engineering. 2018:1-17. DOI: 10.1080/12269328.2018.1515670

[52] Bazazi P, Gates ID, Sanati Nezhad A, Hejazi SH. Silica-based nanofluid heavy oil recovery a microfluidic approach. In: SPE Canada Heavy Oil Technical Conference. 2017. pp. 1-15. DOI: 10.2118/185008-MS

[53] Vatanparast H, Javadi A, Bahramian A. Silica nanoparticles cationic surfactants interaction in water-oil system. Colloids and Surfaces A: Physicochemical and Engineering Aspects. 2017;**521**:221-230. DOI: 10.1016/j.colsurfa.2016.10.004

[54] Zargartalebi M, Barati N, Kharrat R. Influences of hydrophilic and hydrophobic silica nanoparticles on anionic surfactant properties: Interfacial and adsorption behaviors. Journal of Petroleum Science and Engineering. 2014;**119**:36-43. DOI: 10.1016/j.petrol.2014.04.010

[55] Suleimanov BA, Ismailov FS, Veliyev EF. Nanofluid for enhanced oil recovery. Journal of Petroleum Science and Engineering. 2011;**78**:431-437. DOI: 10.1016/j.petrol.2011.06.014

[56] Ahmadi MA, Shadizadeh SR. Induced effect of adding nano silica on adsorption of a natural surfactant onto sandstone rock: Experimental and theoretical study. Journal of Petroleum

Science and Engineering. 2013;**112**: 239-247. DOI: 10.1016/j.petrol.2013.11.010

[57] Sharma T, Iglauer S, Sangwai JS. Silica Nanofluids in an oilfield polymer polyacrylamide: Interfacial properties, wettability alteration, and applications for chemical enhanced oil recovery. Industrial and Engineering Chemistry Research. 2016;**55**:12387-12397. DOI: 10.1021/acs.iecr.6b03299

[58] Bayat AE, Rajaei K, Junin R. Assessing the effects of nanoparticle type and concentration on the stability of CO_2 foams and the performance in enhanced oil recovery. Colloids and Surfaces A: Physicochemical and Engineering Aspects. 2016;**511**:222-231. DOI: 10.1016/j.colsurfa.2016.09.083

[59] Espinoza DA, Caldelas FM, Johnston KP, Bryant SL, Huh C. Nanoparticle-stabilized supercritical CO_2 foams for potential mobility control applications. In: SPE-129925-MS, Present. SPE Improv. Oil Recover. Symp. 24–28 April. Tulsa, Oklahoma, USA, Society of Petroleum Engineers. 2010. pp. 1-13. DOI: 10.2118/129925-MS

[60] Yu J, Liu N, Li L, Lee RL. Generation of nanoparticle-stabilized supercritical CO_2 foams. In: C. Pap. Present. Carbon Manag. Technol. Conf. 7–9 February, Orlando, Florida, USA, Orlando, Florida, USA: Carbon Management Technology Conference. 2012. pp. 1-11. DOI: 10.7122/150849-MS

[61] Yang W, Wang T, Fan Z. Highly stable foam stabilized by alumina nanoparticles for EOR: Effects of sodium Cumenesulfonate and electrolyte concentrations. Energy & Fuels. 2017;**31**:9016-9025. DOI: 10.1021/acs.energyfuels.7b01248

[62] Mo D, Yu J, Liu N, Lee RL. Study of the effect of different factors on nanoparticle-Stablized CO_2 foam for mobility control. In: SPE-159282-MS SPE Annu. Tech. Conf. Exhib. 8–10 October, San Antonio, Texas, USA, Society of Petroleum Engineers. 2012. pp. 1-11. DOI: 10.2118/159282-MS

[63] Worthen A, Bagaria H, Chen Y, Bryant SL, Huh C, Johnston KP. Nanoparticle stabilized carbon dioxide in water foams for enhanced oil recovery. In: SPE-154285-MS, Present. SPE Improv. Oil Recover. Symp. 14–18 April. Tulsa, Oklahoma, USA, Tulsa, Oklahoma, USA: Society of Petroleum Engineers. 2012. pp. 1-7. DOI: 10.2118/154285-MS

[64] Risal AR, Manan MA, Yekeen N, Azli NB, Samin AM, Tan XK. Experimental investigation of enhancement of carbon dioxide foam stability, pore plugging, and oil recovery in the presence of silica nanoparticles. Petroleum Science. 2018;**16**(2):344-356. DOI: 10.1007/s12182-018-0280-8

[65] Kalyanaraman N, Arnold C, Gupta A, Tsau JS, Ghahfarokhi RB. Stability improvement of CO_2 foam for enhanced oil-recovery applications using polyelectrolytes and polyelectrolyte complex nanoparticles. Journal of Applied Polymer Science. 2017;**134**(1–15):44491. DOI: 10.1002/app.44491

[66] Singh R, Gupta A, Mohanty KK, Huh C, Lee D, Cho H. Fly ash nanoparticle-stabilized CO_2-in-water foams for gas mobility control applications. In: SPE-175057-MS, SPE Annu. Tech. Conf. Exhib. 28–30 Sept. Houston, Texas, USA, Houston, Texas, USA: Society of Petroleum Engineers. 2015. pp. 1-13. DOI: 10.2118/175057-MS

[67] Singh R, Mohanty KK. Synergy between nanoparticles and surfactants in stabilizing foams for oil recovery. Energy & Fuels. 2015;**29**:467-479. DOI: 10.1021/ef5015007

[68] Manan MA, Farad S, Piroozian A, Esmail MJA. Effects of nanoparticle

types on carbon dioxide foam flooding in enhanced oil recovery. Petroleum Science and Technology. 2015;**33**: 1286-1294. DOI: 10.1080/10916466.2015.1057593

[69] Eftekhari AA, Krastev R, Farajzadeh R. Foam stabilized by Fly ash nanoparticles for enhancing oil recovery. Industrial and Engineering Chemistry Research. 2015;**54**: 12482-12491. DOI: 10.1021/acs.iecr.5b03955

[70] Kim I, Worthen AJ, Johnston KP, DiCarlo DA, Huh C. Size-dependent properties of silica nanoparticles for Pickering stabilization of emulsions and foams. Journal of Nanoparticle Research. 2016;**18**:82. DOI: 10.1007/s11051-016-3395-0

[71] Singh R, Mohanty KK. Foams stabilized by In-situ surface-activated nanoparticles in bulk and porous media. SPE Journal. 2016;**21**:121-130. DOI: 10.2118/170942-PA

[72] Wang S, Chen C, Kadum MJ, Shiau B-J, Harwell JH. Enhancing foam stability in porous media by applying nanoparticles. Journal of Dispersion Science and Technology. 2018;**39**: 734-743. DOI: 10.1080/01932691.2017.1388175

[73] Singh R, Mohanty KK. Foam flow in a layered, heterogeneous porous medium: A visualization study. Fuel. 2017;**197**:58-69. DOI: 10.1016/j.fuel.2017.02.019

[74] Singh R, Mohanty KK. Nanoparticle-stabilized foams for high-temperature, high-salinity oil reservoirs. In: SPE-187165-MS, Present. SPE Annu. Tech. Conf. Exhib. 9–11 October, San Antonio, Texas, USA, San Antonio, Texas, USA: Society of Petroleum Engineers. 2017. pp. 1-15. DOI: 10.2118/187165-MS

[75] Kumar N, Gaur T, Mandal A. Characterization of SPN Pickering emulsions for application in enhanced oil recovery. Journal of Industrial and Engineering Chemistry. 2017;**54**: 304-315. DOI: 10.1016/j.jiec.2017.06.005

[76] Rezaei A, Abdi-Khangah M, Mohebbi A, Tatar A, Mohammadi AH. Using surface modified clay nanoparticles to improve rheological behavior of Hydrolized Polyacrylamid (HPAM) solution for enhanced oil recovery with polymer flooding. Journal of Molecular Liquids. 2016;**222**: 1148-1156. DOI: 10.1016/j.molliq.2016.08.004

[77] Wei B, Romero-Zerón L, Rodrigue D. Oil displacement mechanisms of viscoelastic polymers in enhanced oil recovery (EOR): A review. Journal of Petroleum Exploration and Production Technologies. 2014;**4**:113-121. DOI: 10.1007/s13202-013-0087-5

[78] Liu P, Mu Z, Wang C, Wang Y. Experimental study of rheological properties and oil displacement efficiency in oilfields for a synthetic Hydrophobically modified polymer. Scientific Reports. 2017;**7**:8791. DOI: 10.1038/s41598-017-09057-9

[79] Kamal MS, Adewunmi AA, Sultan AS, et al. Journal of Nanomaterials. 2017;**2017**:15. DOI: 10.1155/2017/2473175

[80] Hu Z, Haruna M, Gao H, Nourafkan E, Wen D. Rheological properties of partially hydrolyzed polyacrylamide seeded by nanoparticles. Industrial and Engineering Chemistry Research. 2017; **56**:3456-3463. DOI: 10.1021/acs.iecr.6b05036

[81] Maghzi A, Kharrat R, Mohebbi A, Ghazanfari MH. The impact of silica nanoparticles on the performance of polymer solution in presence of salts in polymer flooding for heavy oil recovery. Fuel. 2014;**123**:123-132. DOI: 10.1016/j.fuel.2014.01.017

[82] Maurya NK, Mandal A. Studies on behavior of suspension of silica nanoparticle in aqueous polyacrylamide solution for application in enhanced oil recovery. Petroleum Science and Technology. 2016;**34**:429-436. DOI: 10.1080/10916466.2016.1145693

[83] Giraldo LJ, Giraldo MA, Llanos S, Maya G, Zabala RD, Nassar NN, et al. The effects of SiO_2 nanoparticles on the thermal stability and rheological behavior of hydrolyzed polyacrylamide based polymeric solutions. Journal of Petroleum Science and Engineering. 2017;**159**:841-852. DOI: 10.1016/j.petrol.2017.10.009

[84] Lai N, Guo X, Zhou N, Xu Q. Shear resistance properties of modified Nano-SiO_2/AA/AM copolymer oil displacement agent. Energies. 2016;**9**: 1037. DOI: 10.3390/en9121037

[85] Haruna MA, Pervaiz S, Hu Z, Nourafkan E, Wen D. Improved rheology and high-temperature stability of hydrolyzed polyacrylamide using graphene oxide nanosheet. Journal of Applied Polymer Science. 2019;**0**(1–13):47582. DOI: 10.1002/app.47582

[86] Maurya NK, Kushwaha P, Mandal A. Studies on interfacial and rheological properties of water soluble polymer grafted nanoparticle for application in enhanced oil recovery. Journal of the Taiwan Institute of Chemical Engineers. 2017;**70**:319-330. DOI: 10.1016/j.jtice.2016.10.021

[87] Liu R, Pu W, Sheng JJ, Du D. Star-like hydrophobically associative polyacrylamide for enhanced oil recovery: Comprehensive properties in harsh reservoir conditions. Journal of the Taiwan Institute of Chemical Engineers. 2017;**80**:639-649. DOI: 10.1016/j.jtice.2017.08.043

[88] Ponnapati R, Karazincir O, Dao E, Ng R, Mohanty KK, Krishnamoorti R. Polymer-functionalized nanoparticles for improving Waterflood sweep efficiency: Characterization and transport properties. Industrial and Engineering Chemistry Research. 2011; **50**:13030-13036. DOI: 10.1021/ie2019257

[89] Cheng Y, Zhao M, Zheng C, Guo S, Li X, Zhang Z. Water-dispersible reactive Nanosilica and poly(2-acrylamido-2-methyl-1-propanesulfonic acid sodium) Nanohybrid as potential oil displacement agent for enhanced oil recovery. Energy & Fuels. 2017;**31**: 6345-6351. DOI: 10.1021/acs.energyfuels.7b00743

[90] Behzadi A, Mohammadi A. Environmentally responsive surface-modified silica nanoparticles for enhanced oil recovery. Journal of Nanoparticle Research. 2016;**18**:1-19. DOI: 10.1007/s11051-016-3580-1

[91] Liu R, Pu W-F, Du D-J. Synthesis and characterization of core–shell associative polymer that prepared by oilfield formation water for chemical flooding. Journal of Industrial and Engineering Chemistry. 2017;**46**:80-90. DOI: 10.1016/j.jiec.2016.10.018

[92] Ye Z, Qin X, Lai N, Peng Q, Li X, Li C. Synthesis and performance of an acrylamide copolymer containing Nano-SiO_2 as enhanced oil recovery chemical. Journal of Chemistry. 2013;**2013**:1-33

[93] Lai N, Wu T, Ye Z, Zhang Y, Zhou N, Zeng F. Hybrid hyperbranched polymer based on modified Nano-SiO_2 for enhanced oil recovery. Chemistry Letters. 2016;**45**:1189-1191. DOI: 10.1246/cl.160554

[94] Maghzi A, Mohebbi A, Kharrat R, Ghazanfari MH. Pore-scale monitoring of wettability alteration by silica nanoparticles during polymer flooding to heavy oil in a five-spot glass micromodel. Transport in Porous Media. 2011;**87**:653-664. DOI: 10.1007/s11242-010-9696-3

[95] Yousefvand H, Jafari A. Enhanced oil recovery using polymer/nanosilica. Procedia Materials Science. 2015;**11**: 565-570. DOI: 10.1016/j.mspro.2015.11.068

[96] Foster EL, Xue Z, Roach CM, Larsen ES, Bielawski CW, Johnston KP. Iron oxide nanoparticles grafted with sulfonated and Zwitterionic polymers: High stability and low adsorption in extreme aqueous environments. ACS Macro Letters. 2014;**3**:867-871. DOI: 10.1021/mz5004213

[97] Cheraghian G, Khalili Nezhad SS, Kamari M, Hemmati M, Masihi M, Bazgir S. Adsorption polymer on reservoir rock and role of the nanoparticles, clay and SiO_2. International Nano Letters. 2014;**4**:114. DOI: 10.1007/s40089-014-0114-7

Chapter 3

Surfactant Flooding for EOR Using Sodium Lignosulfonate Synthesized from Bagasse

Rini Setiati, Septoratno Siregar, Taufan Marhaendrajana and Deana Wahyuningrum

Abstract

Surfactant injection is one of the types of chemical injections used in enhanced oil recovery (EOR) process. Surfactant can increase the interfacial tension between oil and water in the rock matrix. The surfactant used is an anionic surfactant, which is one of the lignosulfonate surfactants known as sodium lignosulfonate (SLS) surfactant derived from lignin. Bagasse is one of the raw materials having a high content of lignin (24–25%). The synthesized bagasse becomes lignosulfonate used as an isolation of lignin and transformed into sodium lignosulfonate by sulfonation process. Based on the characteristic test, the bagasse's SLS surfactant has some qualified characteristics which are a good aqueous stability, clear solution and not causing turbidity, and capability to form middle-phase microemulsion with light oil. Synthesized SLS has a hydrophilic–lipophilic balance (HLB) value of 11.6 which can be classified as oil in water (O/W) emulsion. Middle-phase emulsion as a characteristic SLS surfactant affects the performance of the SLS surfactant injection. So the use of sodium lignosulfonate surfactant synthesized from bagasse is a challenge to be developed further as a surfactant flooding.

Keywords: adsorption, bagasse, enhanced oil recovery, light oil, middle-phase emulsion, recovery factor, sodium lignosulfonate surfactant

1. Introduction

Enhanced oil recovery is the process of increasing oil recovery by injecting liquid into a reservoir where the fluid can change the physical properties of oil that is still trapped inside it. One method used is chemical injection, using an anionic surfactant. Surfactants have the ability to increase the interfacial tension between oil and water in the rock matrix. The type of surfactant that is used for the EOR is lignosulfonate surfactant known as sodium lignosulfonate surfactant derived from lignin. One of the raw materials having a high content of lignin is bagasse, which contains 24–25% lignin.

The bagasse process becomes lignosulfonate becoming lignosulfonate uses isolation of lignin from bagasse, and the sulfonated lignin becomes sodium lignosulfonate. For sulfonation process sodium bisulfite ($NaHSO_3$) as reagent can be used. This sulfonation aims to change the hydrophilicity of less polar lignin (insoluble in water) by substituting hydroxyl groups with sulfonate groups so that hydrophility increases [1].

In this study, lignin has been successfully isolated from dried bagasse, which was confirmed through the analysis results of Fourier-transform infrared (FTIR). The FTIR showed the spectrum of isolated lignin has the same characteristic within lignin standard, which contains substituted phenolic aromatic groups and aliphatic alkene groups [2]. Sodium lignosulfonate (SLS) surfactant from bagasse has the same characteristics within the sodium lignosulfonate standard, which comprises the absorption peaks corresponding with the stretching vibration of aromatic and aliphatic alkene functional groups, the stretching vibration of sulfonate groups, and the bending vibration of carboxylic groups [3].

Surfactants will spread in water and absorb at the interfacial tension between air and water or at the interfacial between oil and water. The surfactant is an amphiphilic organic compound, which contains a hydrophobic tail and a hydrophilic head. The aggregate shape may depend on the chemical structure of the surfactant, taking into account the balance of hydrophobic tail and hydrophilic head, known as HLB [4]. Hydrophilic and hydrophobic functions are to determine the emulsion of the system. The type of emulsion can be determined using the hydrophilic–lipophilic balance (HLB) test [5]. The value of HLB shows the tendency of surfactants to dissolve in water or oil to form O/W or W/O emulsion types. Low HLB indicates that the surfactant is more soluble in water, named W/O emulsion type. If the formation salinity is low, low HLB surfactant must be chosen, such that surfactants can make intermediate-phase microemulsions with low salinity. When the formation salinity is high, high HLB makes intermediate-phase microemulsions with high salinity. HLB is determined by calculating values for various molecular regions, giving results on a scale ranging from 0 to 20, for nonionic surfactants. The HLB value 0 corresponds to the total hydrophobic molecule, and the value 20 corresponds to a molecule consisting of a hydrophilic component. HLB values can be used to estimate the nature of surfactants. HLB calculations [6] can be determined by the following formula:

$$HLB = 20 * (M_h)/(M_l + M_h) \tag{1}$$

M_h = molecular weight of hydrophilic groups.

M_l = molecular weight of a hydrophobic or lipophilic group.

For surfactant classification as emulsifier in the O/W system, the surfactant should have a range of values between 8 and 18 **Tables 1** and **2** [6].

The definition of emulsion is a two-phase system in which one of the fluids is dispersed in another, in the form of small droplets [7]. There are two types of emulsions based on the type of liquid that function as internal or external phases, namely:

a. Oil in water (O/W)-type emulsion, which is an emulsion consisting of oil grains scattered into the water

b. Water in oil (W/O)-type emulsion, which is an emulsion consisting of water droplets which are dispersed into oil

Figure 1 explains about W/O emulsion or O/W emulsion. For W/O emulsion, the water is an internal phase and oil is an external phase. As for the O/W emulsion, the oil is an internal phase and water is an external phase.

Microemulsion can occur due to the equilibrium between hydrophiles (head) and lipophiles (tail) so that surface-active agents (surfactants) are formed and reduce interfacial tension. In systems with the same conditions between the number C belonging to crude oil and the surfactant components, there will be a balance between oil and water so that the hydrophilic component (head) of the surfactant

HLB range	General application
2–6	W/O emulsification
7–9	Wetting and spreading
8–18	O/W emulsification
3–15	Detergency
15–18	Solubilization

Table 1.
HLB value and its application [6].

Application	Range HLB
Defoaming of aqueous systems	1–3
detergency and cleaning	12–15
W/O emulsification	3–6
O/W emulsification	8–28
Solubilization	11–18
Wetting	7–9

Table 2.
Required HLB for a number of applications [4].

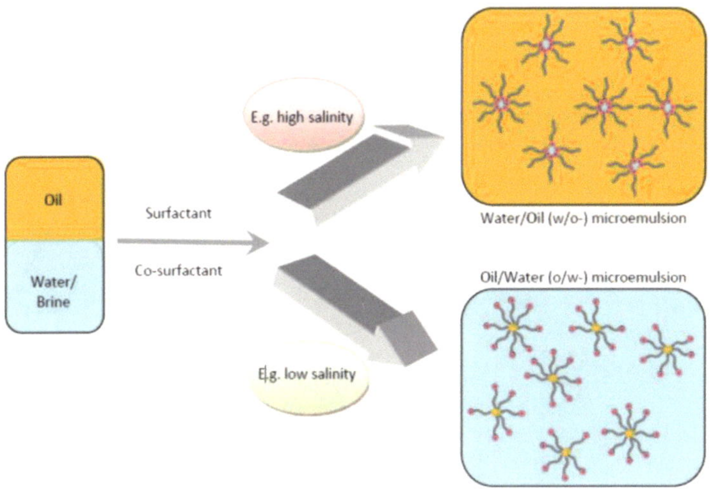

Figure 1.
Oil and water formation forming microemulsion [8].

will go to the surface and form a surface-active agent (surfactant). In a high-salinity oil-soluble surfactant system, there will be ionic strength that is identical to salinity where the hydrophilic (head) component of the surfactant is weak and the lipophile (tail) component is strong. So, in a system with high salinity, the hydrophilic component weakens, and the lipophilic component will be stronger so that a balance occurs in the system and a surface-active agent (surfactant) is formed. At high salinity, the salt component will make a balance between the components C of surfactant and oil, thereby strengthening the interaction between surfactants and oil. Salinity of the formation water affects the reduction of oil–water interfacial tension by surfactant.

The emulsion condition formed can be seen from the results of the phase behavior test, which is a test of mixing fluid reservoir with injection fluid.

Emulsions are made to obtain stable and flat preparations from mixtures which cannot mix together. The behavior of the mixed phase is strongly influenced by solvent water salinity, temperature, type and concentration of surfactant, and oil type. The type of emulsion is also determined by salinity, as shown in **Figure 2**. In the figure, the optimal emulsion is in the middle, where the emulsion occurs in the middle area between water (surfactant) and oil, called the middle-phase emulsion.

Observations of mixed-phase behavior between reservoir fluid and injection fluid can be classified as lower-phase emulsion, microemulsion (middle-phase emulsion), upper-phase emulsion, macroemulsion, and sediment. **Figure 3** can explain about the classification of emulsion. This test is carried out on surfactants dissolved in formation water and then mixed with crude oil. This test aims to get the right conditions for the surfactant by determining the exact salinity and type of emulsions that can mix with crude oil.

The emulsion formed indicates that in the system there is a decrease in interfacial tension between the surfactant and crude oil systems. This phase behavior test is strongly influenced by the salinity of formation water.

Increased salinity will reduce the solubility of anionic surfactants in formation water, because with increasing concentration, the surfactant electrolyte is released from the formation water system [9]. In the middle phase that forms microemulsions, it represents the condition of miscible displacement, whereas in the mixed phase which forms the upper-phase or lower-phase emulsion, it represents an immiscible displacement condition.

The emulsion formed can be an indicator of the IFT value of the system. A high IFT indicates an immiscible fluid, while a low IFT indicates a fluid is miscible.

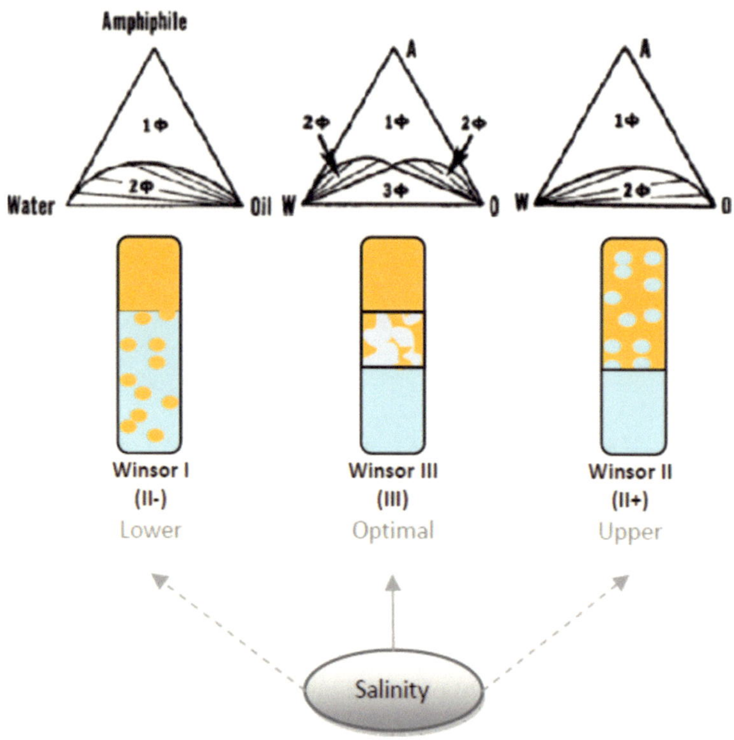

Figure 2.
Emulsion type based on salinity [8].

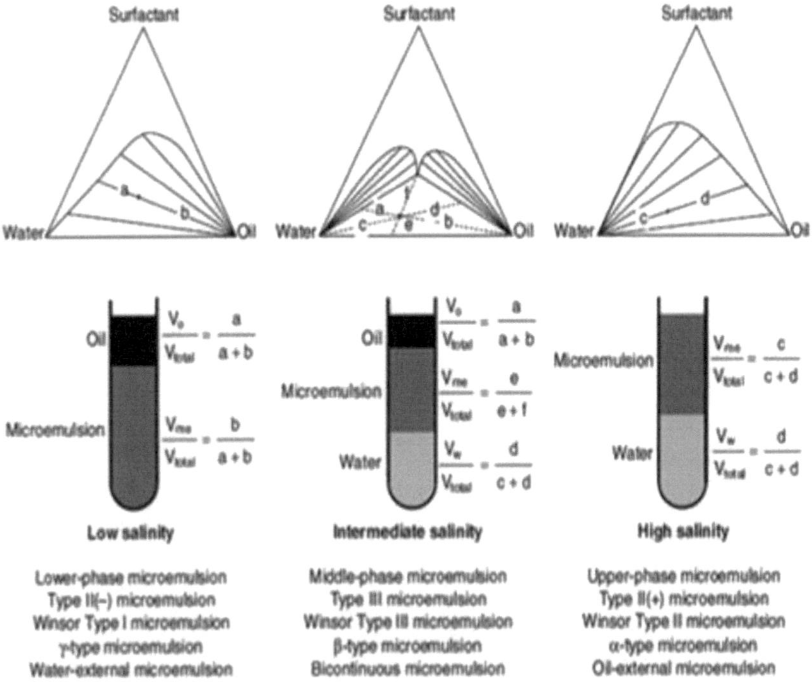

Figure 3.
Three types of microemulsion effect of salinity and phase behavior [5].

IFT at surfactant and oil is a function of formation water salinity, temperature, surfactant concentration, type of surfactant, and oil type, while the phase behavior test is influenced by the type of oil, type of co-surfactant, temperature, structure of surfactant, and pressure [5]. The behavior of the microemulsion phase is very complex and depends on a number of parameters. Surfactants can form microemulsions because of their level of solubility that are both in water and in oil. There is no equation for simple microemulsions. Therefore, the phase behavior for certain microemulsion systems must be measured experimentally. The phase behavior of microemulsions is usually presented using ternary diagrams and empirical correlations. The ternary diagram is a very useful tool in EOR because it can simultaneously represent the composition of phases and the whole and relative numbers. **Figure 4** shows a ternary diagram schematic.

In the ternary diagram, the equilateral triangle represents the components of water, oil, and 100% surfactant from the solution. The concentration can be shown in mole fraction, mass, or volume. The single-phase region is in the zone of high surfactant concentration. The three-phase region is in the middle zone. The two-phase lobes (vertices) are in the upper right and left over the three-phase triangle. There is a third two-phase region which is located at very low surfactant concentrations below the three-phase region. Type III ternary-phase behavior consists of an area close to the axis of salt water/oil which is bounded by a triangle. The composition in this area will produce three phases; the composition of each phase is the same as the composition of the curved triangular peak.

The microemulsion system can also be designed to obtain a very low IFT value, around 0.001 mN/m by making a stable brine solution or with a hydrocarbon phase [10] which will be advantageous for the EOR process. In the EOR process, an important part of the ternary diagram is the three-phase region. The general form of ternary diagram can be classified as type II (−), namely, lower-phase emulsion

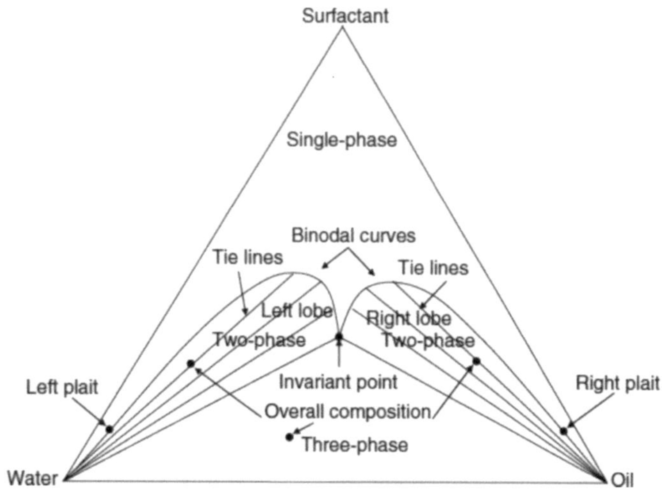

Figure 4.
Ternary diagram schematic [5].

and excess oil phase; type (+), namely, upper-phase emulsion with excess water phase; and type III, middle-phase microemulsion. The lowest interfacial tension occurs because the SLS surfactant can form a microemulsion.

The SLS surfactant from bagasse can function as an oil in water emulsion which means that the SLS surfactant is soluble in water but does not mix. For the certain salinity, NaCl will cause an effective reduction of the oil–water interfacial tension. This condition occurs in the proportion of 20,000 ppm, 4.5%; 40,000 ppm, 1.5%; and 40,000 ppm, 4%, where the increase in salt (NaCl) causes a decrease of interfacial tension and reduces oil recovery. In the proportion of salinity of 20,000 ppm, 4.5%, IFT is lower, but microemulsions are not formed, meaning that in this system the surfactant is not soluble in the oil. This condition causes the reduction of performance of bagasse SLS surfactant and produces less recovery factor of oil [11].

2. Methodology

Bagasse is a by-product of the sugarcane liquid extraction process. Bagasse is one of the sources of biomass, whose utilization is currently mostly only as fuel in steam boilers, as raw materials of paper, or as a source of animal feed. The bagasse used is bagasse which has undergone the fifth milling process from the process of sugar (**Figure 5**) [11].

Bagasse mostly contains lignocellulose. The fiber length is between 1.7 and 2 mm with a diameter of about 20 μm. Bagasse contains 48–52% water, average sugar is 3.3%, and average fiber is 47.7%. Bagasse fiber is insoluble in water and consists mainly of cellulose, pentose, and lignin. Based on chemical analysis, the average bagasse has a chemical composition, namely, ash 3.28%, lignin 22.09%, cellulose 37.65%, juice 1.81%, pentosan 27.97%, and SiO_2 3.01% [6].

The process of bagasse becomes lignosulfonate starting from the separation of lignin from bagasse and then reaction with sodium bisulfite to sodium lignosulfonate. The processing of bagasse to produce lignosulfonate was initially performed by isolation of lignin and the isolated lignin transformed into sodium lignosulfonate by sulfonation process. The process is using reflux equipment with a condenser and bagasse at 80 mesh in size. The lignin isolation process uses reflux for 4 h, with sodium hydroxide reagent. The lignin produced was then processed into lignosulfonate using sodium bisulfite reagent and refluxed for 4 h [11].

Figure 5.
Sugar cane—Bagasse [11].

No.	Surfactant composition
1	5000 ppm—1.5%
2	10,000 ppm—1.5%
3	10,000 ppm—3.0%
4	20,000 ppm—1.5%
5	20,000 ppm—4.5%
6	40,000 ppm—1.5%
7	40,000 ppm—4.0%
8	80,000 ppm—1.5%
9	80,000 ppm—4.5%

Table 3.
Surfactant composition of concentration and salinity [11].

Before use as laboratory-scale injection fluid, the SLS surfactant synthesized from bagasse was tested for its characteristics to match the fluid and reservoir rocks to be injected. A variation of surfactant concentration and variation of formation water salinity were used. The SLS surfactant synthesized from bagasse is injected into a 20% porosity and 100–500 mD Berea cores containing synthetic brine and light crude oil. The injection process with bagasse SLS surfactant has been carried out with variations in various proportions of salinity and surfactant concentration, as stated in **Table 3**.

The core injection process uses a surfactant solution of 1.5, 3.0, 4.0, and 4.5% with salinity of 5000, 10,000, 20,000, 40,000, and 80,000 ppm. The core injection uses core flood apparatus with Berea core synthesis in the core holder, shown in **Figure 6**. The core injection process consists of two stages: water injection and sodium lignosulfonate surfactant injection [11].

Before surfactant injection, cores were saturated with formation water using a vacuum system, until 100% saturated. Formation water is injected into the core with a variation of the flow rate, 0.5, 0.75, 1.0, and 1.5 cc/min, so that the core is filled with formation water. From this stage, effective water permeability (K_w) data will be obtained. In the next stage, crude oil is injected into the core with a variation of flow rate, 0.25, 0.5, 1.0, 1.5, and 2.0 cc/min, so the formation water is pushed out, and the core is filled with crude oil. From this stage, effective oil permeability (K_o) and connate water saturation (S_{wc}) can be measured.

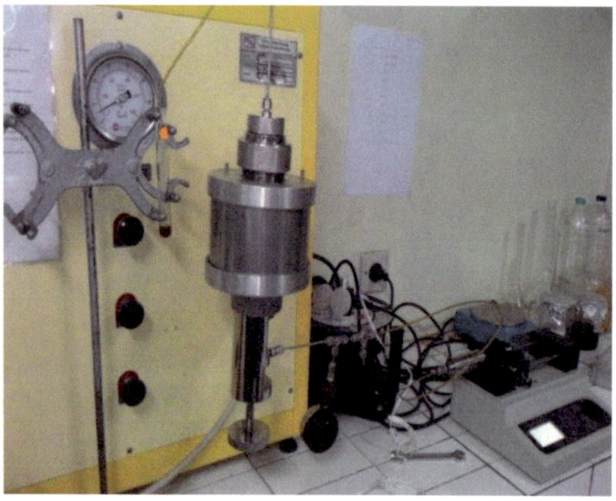

Figure 6.
Core flood apparatus [11].

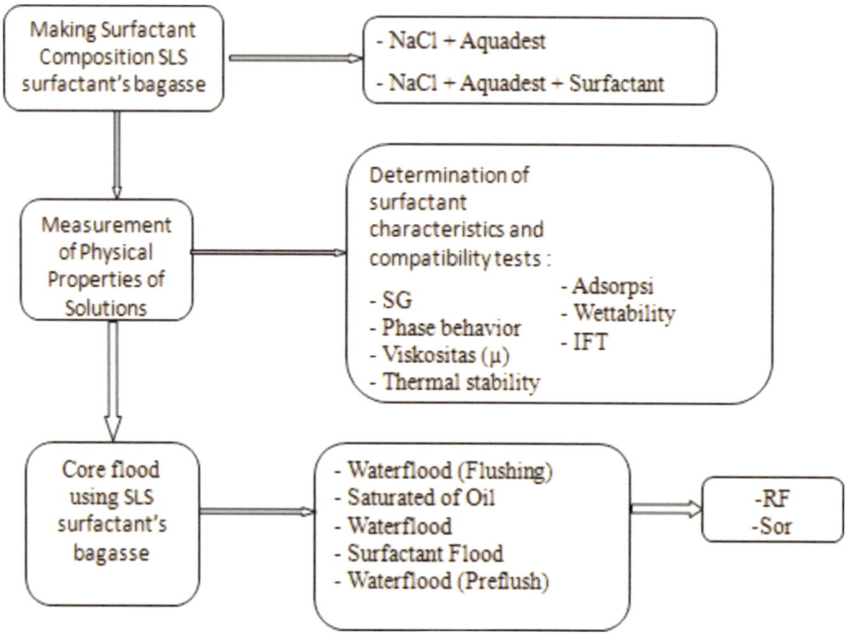

Figure 7.
Scheme of surfactant injection using SLS surfactant synthesized from bagasse [11].

After the core is saturated with crude oil, the first process begins with water injection and then proceeds with injection of surfactant, as shown below (**Figure 7**).

3. Result and discussion

The analysis of nuclear magnetic resonance (NMR) spectroscopy of SLS surfactant showed that the surfactant consisted of monomer structure having methoxyl and hydroxyl substituted benzene as well as the presence of sulfonate and aliphatic hydroxyl groups. Therefore, according to NMR spectrum analysis, it can be deduced that the monomer of synthesized surfactant has empirical formula of $(C_{11}H_{16}O_8S)_n$ with relative molecular weight of 308.06. The exact molecular weight

No.	Functional groups in structure Lignosulfonate	Wave number (cm^{-1})		
		SLS surfactant synthesized from bagasse	Lignosulfonate surfactant (Aldrich)	Lignosulfonate surfactant (Patricia)
1.	Alkene —C=C—	1635.34	1608.34	1630–1680
2.	Sulfonate S=O	1384.64	1365	1350
3.	Carboxylic C=O	1114.64	1187.94	1000–1300
4.	Ester S—OR	462.832	499.831	500–540

Table 4.
Results of FTIR of SLS surfactant synthesized from bagasse [2].

of the synthesized surfactant should be further determined using mass spectrometry measurement. The monomer of the synthesized SLS has a hydrophilic–lipophilic balance value of 11.6 which can be classified as O/W emulsion [12, 13], which means that the SLS surfactant is water soluble. Thus, the SLS surfactant derived from bagasse can be used as an injection fluid and formed middle-phase emulsion that is required in surfactant injection performance.

The results of the Fourier-transform infrared test on bagasse SLS surfactant have shown the existence of components contained in the SLS surfactant which consists of four main components forming lignosulfonic surfactants: alkene stretching groups, sulfonate stretch groups, carboxylic buckling groups, and ester buckling groups with wave numbers from the results of the FTIR [2]. As shown in **Table 4**, it turns out that the wave numbers in the SLS bagasse surfactant component have similarities with the same component wave number for standard lignosulfonic surfactants, namely, standard lignosulfonates from Aldrich and Patricia [14].

Before the characteristics of SLS surfactant synthesized from bagasse was carried out, the physical properties of the SLS surfactant were first measured. The physical properties of the SLS surfactant measured included the viscosity, density, and pH of the SLS surfactant at various proportions of the composition to be used in the injection. The measurement results can be seen in **Table 5**.

From the measurement of physical properties, it turns out SLS surfactant solution in various proportions has physical properties that are almost the same as the viscosity data range 1.5350–1.8375 cP, density 0.9970–1.0657 g/cc, and pH 5.91–6.67. The physical properties of this surfactant are in accordance with the characteristics of the surfactant as described previously.

The characteristic test of the bagasse SLS surfactant consists of:

a. Aqueous stability test.

b. Phase behavior test/emulsion test.

c. Interfacial tension (IFT) test.

d. Thermal stability test.

e. Adsorption test.

f. Wettability test.

From the characteristic test, some of the properties of SLS surfactant from bagasse can be known as seen in **Table 6**.

No.	Surfactant composition	Viscosity (cP)	Density (g/cc)	pH
1	10,000 ppm—1.5%	1.5350	1.0003	6.39
2	10,000 ppm—3.0%	1.5443	1.0113	6.67
3	20,000 ppm—1.5%	1.6877	1.0069	5.91
4	20,000 ppm—4.5%	1.5757	1.0280	6.56
5	40,000 ppm—1.5%	1.6697	1.0189	6.12
6	40,000 ppm—4.0%	1.7724	1.0367	6.46
7	80,000 ppm—1.5%	1.7630	1.0438	5.86
8	80,000 ppm—4.5%	1.8375	1.0657	4.38

Table 5.
The physical properties of the SLS surfactant synthesized from bagasse [11].

No.	Salinity (ppm)	Surfactant concentration (%)	Aqueous stability	Middle emulsion (Stability) %	IFT (mN/m)
1.	10,000	1.5	Clear	10.00	2.73
2.	10,000	3.0	Clear	7.50	1.68
3.	20,000	1.5	Clear	5.00	4.13
4.	20,000	4.5	Clear	0.00	1.27
5.	40,000	1.5	Clear	6.00	4.11
6.	40,000	4.0	Clear	0.00	2.72
7.	80,000	1.5	Clear	1.25	3.61
8.	80,000	4.5	Clear	1.00	1.09

Table 6.
Characteristic test results in various variations of SLS surfactant concentrations [11].

No.	Salinity (ppm)	Concentration surfactant (%)	IFT (mN/m)	Recovery factor from surfactant flooding (%)
1	5000	1.5	6.81	7.00
2	10,000	1.5	2.73	9.25
3	20,000	1.5	4.13	8.55
4	40,000	1.5	4.11	1.80
5	80,000	1.5	3.61	10.71

Table 7.
IFT and contact angle to recovery oil by SLS surfactant injection of bagasse (concentration of surfactant 1.5%).

In this research, the surfactant composition of 4.5% of SLS surfactant in 80,000 ppm NaCl gave the lowest IFT value of 1.091 mN/m. From the characteristic test, SLS surfactant synthesized from bagasse has aqueous stability, is clear, and does not cause turbidity. It has a hydrophilic–lipophilic balance value of 11.6 which is classified as oil in water emulsion and can form microemulsion with light oil. This SLS surfactant has middle-phase emulsion with volume fraction of 1–10% microemulsion and IFT of 1.091–6.81 mN/m.

Surfactant injection processes in this research using a variety of concentrations using a variety of concentrations and salinities have shown in **Table 7** and **Figure 8** the contour of relation between salinity and IFT to recovery factor.

From **Table 7**, optimal enhanced oil recovery reaches 10.71% at a concentration of surfactant of 1.5% and 80,000 ppm of surfactant solution. From the oil recovery data, the salinity proportion of 10,000 ppm of 1.5% oil recovery with surfactant injection (RF SF) reached 9.25% with stable middle-phase emulsion up to 10% and IFT value of 2.73 mN/m (**Table 6**). This is consistent with the theory that surfactants serve to lower IFT causing a breakdown of the water-to-water interfacial tension resulting in emulsions being formed up to 10% because they are soluble in oil and water. This microemulsions cause the SLS surfactant to produce the lowest value interfacial tension IFT 2.73 mN/m [15].

From **Figure 8**, in the relation contour between salinity and IFT, there are two areas: high recovery factor (red–orange color) and low recovery factor (purple–blue color). **Figure 8** and **Table 7** explain that at low salinity, 10,000 ppm (1.5% surfactant), the recovery factor is 9.25% and at high salinity, 80,000 ppm (1.5% surfactant), the recovery factor is 10.71%. These are the highest recovery factors at this surfactant injection using SLS surfactant synthesized from bagasse. The lowest recovery factor occurs at 40,000 ppm, only 1.80%. The results of injection of core using SLS surfactant from bagasse at concentration of 1.5% can be seen in **Table 7**.

Furthermore, the surfactant with the optimum composition has been tested toward its static and dynamic adsorptions. At the same surfactant concentration in different salt contents, it showed that the higher the salt content, the lower the adsorption value, which means that the surfactant is less absorbed into the rock when the surfactant was injected to the core. Since the less surfactant was

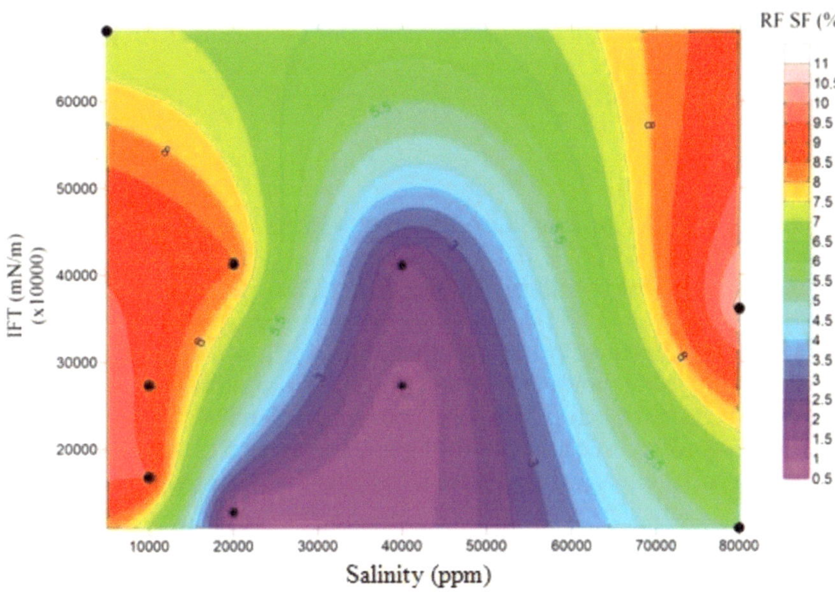

Figure 8.
Relation contour between salinity and IFT to recovery factor [11].

No.	Surfactant composition	Static adsoprtion (%)	Dynamic adsorption (%)
1.	20,000 ppm—4.5%	20.533	29.16
2.	80,000 ppm—1.5%	8.071	1.37

Table 8.
The result of static and dynamic adsorption test [15].

absorbed into the core, it means that more surfactants were available and capable to decrease the interfacial tension between oil and water; therefore the more oil can be produced by the surfactant injection. Wettability test results also showed that the higher salt content produces a larger contact angle which means the system becomes more water wet. These conditions have made the synthesized SLS surfactant derived from bagasse more dissolved in formation water to release grains of oil attached to the core. From the injection process, the composition of surfactant, with a salinity of 1.5% (80,000 ppm), showed the highest oil recovery value up to 10.71%, compared to the other composition. Therefore, the surfactant concentration and salinity affected many factors related to the performance of SLS surfactant of bagasse displacement on light oil.

At the low area recovery factor, the composition of surfactant, with a salinity of 1.5% (40,000 ppm), showed the lowest oil recovery value of 1.80%. At this condition, the value of IFT is 4.11 (mN/m). This value can explain that for this area, they are insoluble for oil and water, so that oil does not move easily. Besides IFT, there are other factors that affect the mechanism surfactant during the surfactant injection process. The other factor is adsorption. **Table 8** indicates the static and dynamic adsorptions of this SLS surfactant on the core surface at the high and the low salinities.

At this salinity of 20,000 ppm, the static adsorption is 20.533%, and the dynamic adsorption is 29.16%. This adsorption is higher than the adsorption at 80,000 ppm. From **Table 7**, recovery factor of the higher adsorption (salinity 40,000 ppm) is 1.80%, and for the lower adsorption (salinity 80,000 ppm), recovery factor is 10.71%. Due to the large amount of adsorption that occurs on the core, the amount of surfactant decreases, so that the surfactant mechanism also decreases. The lower mechanism of surfactant results in a decrease of recovery factor.

4. Conclusion

The SLS surfactant synthesized from bagasse with its natural characteristic can be used as an injection fluid in surfactant injection to increase oil recovery for sandstone with low to high salinity reservoir. The SLS surfactant synthesized from bagasse has displaced some of the residual oil after waterflood, depending on the IFT reduction, middle-phase emulsion (microemulsion), and adsorption value. The results of this study showed that the synthesized SLS surfactant of bagasse with its natural characteristics can be used as injection fluid in surfactant injection to increase oil recovery, especially for the light oil recovery. So the use of sodium lignosulfonate surfactant synthesized from bagasse is a challenge to be developed further as a surfactant flooding.

Acknowledgements

Thanks to the OGRINDO Consortium Research at the Bandung Institute of Technology (ITB), Trisakti University, and the Indonesian Directorate General of Higher Education Decentralization Fund for PUPT Fund 2015–2018 ITB for facilitating this research. We also thank IntechOpen who published this article, which in turn will benefit the society.

Author details

Rini Setiati[1]*, Septoratno Siregar[2], Taufan Marhaendrajana[2] and Deana Wahyuningrum[3]

1 Department of Petroleum Engineering, Faculty of Earth Technology and Energy, Universitas Trisakti, Jakarta, Indonesia

2 Faculty of Mining Petroleum Engineering, Faculty of Mining and Petroleum Engineering, Institut Teknologi Bandung, Bandung, Indonesia

3 Department of Chemistry, Faculty of Math and Science, Institut Teknologi Bandung, Bandung, Indonesia

*Address all correspondence to: rinisetiatidgm@gmail.com

References

[1] Sheng JL. Chemical flooding. In: Modern Chemical Enhanced Oil Recovery. Oxford: Elsevier; 2011. pp. 239-335

[2] Setiati R, Putri EAM, Wahyuningrum D, Siregar S, Marhaendrajana T, Fajriah S. Sulfonation of Lignin's bagasse into lignosulfonate surfactants as alternative raw materials for injection of surfactants in the petroleum industry. In: Proceedings of the Lignocellulose Seminar; Center for Biomaterial Research LIPI; 2016. Available from: www.biomaterial.lipi.go.id/epub/index.php/1/prosligno

[3] Sandersen SB. Enhanced Oil Recovery with Surfactant Flooding. Denmark: Center for Energy Resources Engineering—CERE; 2012

[4] Myers D. Surfactant Science and Technology. New Jersey: Wiley Interscience; 2006

[5] Setiati R, Siregar S, Marhaendrajana T, Wahyuningrum D, Synergetic Effect of SLS Surfactant of Bagasse to Enhanced Oil Recovery. In: Proceedings of the 1st Conference of the Arabian Journal of Geosciences (CAJG-1); March 2019. Springer International Publishing; 2019. DOI: 10.1007/978-3-030-01578-7_17

[6] Hermiati E. Utilization of lignocellulosic biomass from sugarcane bagasse for bioethanol. Journal of Agricultural Research and Development. 2010;**29**(4):121-130. DOI: 10.21082/jp3.v29n4.2010

[7] Acosta E. Modeling and Formulation of Microemulsions: The Net-Average Curvatre Model and the Combine Linker Effect. Norman: University of Oklahoma; 2004

[8] Schramm L. Surfactant: Fundamentals and Application in the Petroleum Industry. Cambrigde: University of Cambrige; 2000

[9] Patricia RJ. Relationship between the structure of Fe-lignosulfonate complexes determined by FTIR spectroscopy and their reduction by the leaf Fe reductase. In: The Proceedings of the International Plant Nutrition Colloquium XVI. Davis: University of California; 2009. p. 1

[10] Green DW, Willhite G. Enhanced Oil Recovery. Texas: SPE; 1998

[11] Setiati R, Siregar S, Marhaendrajana T, Wahyuningrum D. Infra red evaluation of sulfonation surfactant sodium lignosulfonate on bagasse. The Interrnational Journal of Science and Technoledge. 2017;**5**(3):137-142. ISSN: 2321-919X. Available from: www.theijst.com

[12] Setiati R. Synthesis and characterization of sodium lignosulfonate from bagasse: The effects of concentration and salinity toward the performance of oil injection in core [Doctoral dissertation]. Indonesia: Bandung Institute of Technology; 2017

[13] Setiati R, Siregar S, Marhaendrajana T, Wahyuningrum D. Application of sodium lignosulfonate surfactant synthesized bagasse as an alternative surfactant flooding for light oil reservoir considering carbon number. In: Simposium IATMI; 1-3 October 2018; Padang, Indonesia; 2018

[14] Akzo N. HLB and Emulsification. Surface Chemistry LLC. 2011. Available from: http://surface.akzonobel.com

[15] Setiati R, Siregar S, Marhaendrajana T, Wahyuningrum D, et al. IOP Conf. Series: Materials Science and Engineering. 2018;**434**:012083. DOI: 10.1088/1757-899X/434/1/012083

Chapter 4

Direct Gas Thickener

Nasser Mohammed Al Hinai, Matthews Myers, Colin D. Wood and Ali Saeedi

Abstract

Direct gas thickening technique has been developed to control the gas mobility in the miscible gas injection process for enhanced oil recovery. This technique involves increasing the viscosity of the injected gas by adding chemicals that exhibit good solubility in common gasses, such as CO_2 or hydrocarbon (HC) solvents. This chapter presents a review of the latest attempts to thicken CO_2 and/or hydrocarbon gases using various chemical additives, which can be broadly categorised into polymeric, conventional oligomers, and small-molecule self-interacting compounds. In an ideal situation, chemical compounds must be soluble in the dense CO_2 or hydrocarbon solvents and insoluble in both crude oil and brine at reservoir conditions. However, it has been recognised that the use of additives with extraordinary molecular weights for the above purpose would be quite challenging since most of the supercritical fluids are very stable with reduced properties as solvents due to the very low dielectric constant, lack of dipole momentum, and low density. Therefore, one way to attain adequate solubility is to elevate the system pressure and temperature because such conditions give rise to the intermolecular forces between segments or introduce functional groups that undergo self-interacting or intermolecular interactions in the oligomer molecular chains to form a viscosity-enhancing supramolecular network structure in the solution. According to this review, some of the polymers tested to date, such as polydimethylsiloxane, polyfluoroacrylate styrene, and poly(1,1-dihydroperfluorooctyl acrylate), may induce a significant increase of the solvent viscosity at high concentrations. However, the cost and environmental constraints of these materials have made the field application of these thickeners unfeasible. Until now, thickeners composed of small molecules have shown little success to thicken CO_2, because CO_2 is a weak solvent due to its ionic and polar characteristics. However, these thickeners have resulted in promising outcomes when used in light alkane solvents.

Keywords: CO_2, associated gas, NGL, polymer, small molecules, miscible gas injection, thickener, and viscosity

1. Introduction

Over the past 50 years, several research groups have attempted to increase the viscosity of gas-solvents for two purposes. The first purpose is to reduce the gas mobility and improve conformance control for miscible gas injection (MGI). By simply increasing the injected gas viscosity, gas mobility will be reduced, and as a result the sweep efficiency and oil recovery would be improved. The second purpose is to thicken supercritical fracturing fluids to enhance well productivity in tight reservoirs [1–3]. This former application is particularly useful for reservoirs that are sensitive to the typical water-based fluids used for fracturing. Increasing the fluid viscosity results in a more effective fracturing fluid [1]. In addition, at high pressures, viscous fluids would be able to propagate wider fractures by improving the transport of proppant particles and reducing the leak-off of gas into the faces of the fracture [1, 3, 4].

In previous studies, efforts were centred on identifying thickeners for CO_2 and natural gas liquid (NGL) (i.e. ethane, propane, and butane) thickeners. These attempts are based on polymeric and small-molecule candidates as will be reviewed and highlighted in this chapter. The mechanisms behind the thickening of any solvent depend on polymer coil expansion, intermolecular interactions, entanglement, aggregation (affected by the polymer molecular weight distributions), and self-assembly and indirectly through the effect of polymer molecules on nearby solvent molecules [5].

2. Direct carbon dioxide thickeners

2.1 Polymeric thickeners

The use of a polymer thickener is one of the fundamental strategies for increasing CO_2 viscosity [2]. The main advantages of this approach are that the thickening process can enhance CO_2 viscosity over a wide range of temperatures and improve sweep efficiency across the reservoir formation [6]. Although high molecular weight (Mw) polymers (Mw >10^6 g.mol^{-1}) are effective viscosity enhancers at dilute concentrations, it is extremely challenging to dissolve ultrahigh molecular weight in dense CO_2 at reservoir conditions [6, 7]. In the literature, several polymers (Mw: 10^3–10^6 g.mol^{-1}) have been designed and identified that can be dissolved and thicken supercritical CO_2 [1]. However, the pressure required for the dissolution of these polymers is very high in the range of 68.95–275.79 MPa, which is significantly higher than the typical reservoir pressures for CO_2 flooding (MMP, 10.3–27.6 MPa) [1].

The earliest attempts at viscosity enhancers for dense CO_2 were with oil-soluble polymers (e.g. non-polar organic polymers) because CO_2 is known to be a non-polar solvent capable of dissolving certain hydrocarbons and other small molecules quite well [6, 8]. Therefore, it was expected that oil-soluble polymers would be a more likely candidate to dissolve in supercritical CO_2 compared to water-soluble polymers. Heller et al. identified 18 hydrocarbon-type polymers that exhibited encouraging solubility (0.22–10 g/litre) in CO_2 at pressures of 11.7–21.4 MPa and temperatures of 293–331 K [9–14]. Although several polymers showed a slight increase in CO_2 viscosity, none of the studied polymers were capable of enhancing the viscosity of CO_2 significantly to a useful level. This is attributed to low solubility in CO_2 leading to a random polymer coil structure that is not efficient at significantly increasing viscosity. Furthermore, the molecular weight of the polymers found to be soluble in CO_2 was very low. For example, a 1 wt% solution of low molecular weight atactic poly(methyl oxirane) (Mw: 408 g.mol^{-1}) exhibited slight solubility in CO_2 and increased its viscosity by approximately 25% at 301–306 K and 13.7–17.9 MPa [11]. These initial efforts were followed by subsequent attempts to maximise the entropy of mixing between the CO_2 and polymers by using disordered polymers with irregularity in multiple components and side chains that varied in chain length [9]. Thereby, focus was put on poly(α-olefins), such as poly(1-hexene), poly(1-pentene), and poly(1-decene) (P-1-D). Although some achievements were made with some of the evaluated polymers, none of these amorphous polymers were considered to be effective thickeners primarily due to their low solubility in CO_2. In general, it is concluded that the molecular weight of the polymers had to be fairly low (Mw > 1000 $g.mol^{-1}$) and thus less effective at enhancing viscosity to achieve significant solubility in CO_2 [6]. Typically, high and ultrahigh molecular weight polymers are used as effective thickeners. In 2012, Zhang et al. reported that less than 1 wt% solutions of two oligomers (i.e. P-1-D and poly(vinyl ethyl ether) (PVEE)) (**Figure 1**) could increase the viscosity of CO_2 by 13–14-fold at 329 K [15]. Previous research groups found that neither a 1 wt% mixture of PVEE nor P-1-D was capable of enhancing the viscosity in either toluene or CO_2 by more than several percents [2, 6, 8]. Therefore, Zhang et al. findings do not correlate and are inconsistent with the results of other research groups. Most previous studies, reported that for low/high molecular weight polymers, a concentration of 1.5–7 wt% is required to thicken CO_2 albeit at very high pressure [6]. In recent studies, [16, 17] P-1-D has been found to have sufficient solubility in both CO_2 and associated gas (AG) mixtures (at temperatures above 358 K and pressures of 50–55 MPa) to considerably increase gas viscosity. The viscosity enhancement of P-1-D in an AG mixture (25 mol% CO_2) and CO_2 was measured in a capillary viscometer at different pressures (50–55 MPa), 377 K, and varying P-1-D concentrations (1.5–9 wt%). For example, at 5 wt% P-1-D, the CO_2 viscosity falls in the range of 0.14–0.18 cP (1.2–2.6-fold) over the pressure range of 50–55 MPa, while over the same pressure range, the AG mixture viscosity is in the range of 0.126–0.131 cP (~4-fold).

In 1987, a patent published by Bullen and co-workers [18] claimed that CO_2-based fracturing could be improved by adding a small amount of a polycarbonate copolymer (Mw: 20,000–150,000 $g.mol^{-1}$) that was formed via low-temperature reaction of CO_2 with propylene oxide in a homogeneous catalyst (e.g. diethylzinc and/or acetic acid anhydride). This copolymer exhibits dissolution in CO_2 and is capable of increasing its viscosity by threefold at a concentration of 2.5 wt% at 295 K and pressures ranging from 10 to 25 MPa. Furthermore, Sarbu et al. [19] tested the solubility of poly(ether-carbonate) copolymers (derived from propylene oxide and CO_2) in CO_2. They found that this copolymer (Mw: 16000 g/mol) could only be dissolved at 1 wt% at 295 K and 14 MPa. However, there was no significant increase in CO_2 viscosity under these conditions. This calls into question some of the results by Bullen and co-workers as they are much better than later literature [2].

Other researchers have attempted to use entrainers (co-solvents) to improve the solubility of polymers in CO_2 [20] and as such increase the CO_2 viscosity as well as increase the solubility of crude oil components in the CO_2-rich phase [21]. These

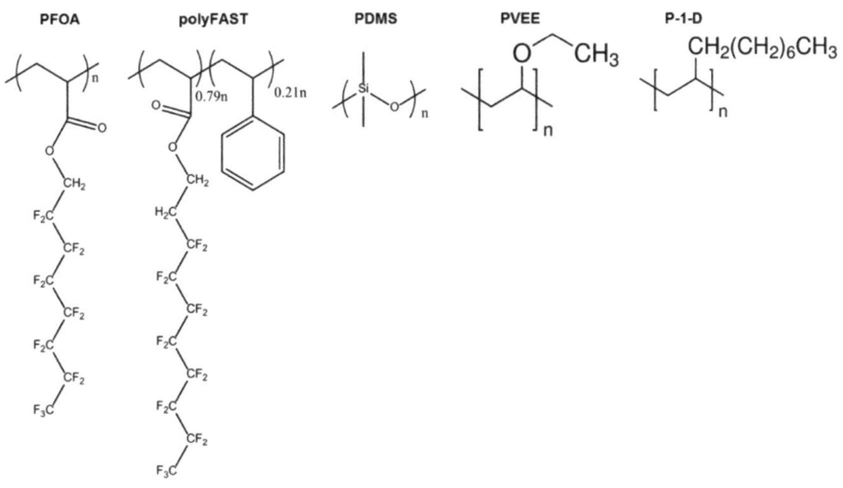

Figure 1.
High/low molecular weight polymeric thickener tested in CO_2 with/without co-solvent [8].

entrainers are relatively low molecular weight non-polar or polar organic compounds which include alcohols, glycols, ethoxylated alcohols, and hydrocarbons [21]. Chullick's patent claimed that addition of alcohol and glycol would promote the solubility of polymers in CO_2 [20]. The rationale is that polar entrainers improve the polarizability of CO_2, which may act in a similar manner to a surfactant in the water/oil system, while non-polar entrainers may function as mutual solvents in a polymer/CO_2 system [20]. Therefore, the addition of entrainers to a supercritical fluid (SCF) leads to increase in the solvent power of SCF [21]. Furthermore, a National Institute for Petroleum and Energy Research (NIPER) group evaluated the use of entrainers (without polymer) as CO_2 thickeners [21]. They reported substantial increase of CO_2 viscosity with high concentration of entrainers in CO_2. For example, 13 mol% of isooctane and 44 mol% of 2-ethylhexanol increased the viscosity of CO_2 by 243 and 1565%, respectively. However, at dilute concentration of entrainers in CO_2, the viscosity enhancement was subtle. For example, 2 mole% of 2-ethylhexanol resulted in 24% of CO_2 viscosity enhancement [21]. In another patent, a treating fluid was used to increase the viscosity of CO_2 solution. This treating fluid is formed by solubilising a polymer or copolymer of dimethylacrylamide (0.25–2.5 wt%) in the substantially anhydrous liquid which was cross-linked by a metal ion source (0.01–2 wt% of titanium, zirconium, and/or aluminium). The substantially anhydrous fluid/polymer and CO_2 solutions formed a single phase and viscosified the fluid (13–30 cP) at temperatures of 338–377 K and pressures of 6.89–12 MPa [22]. Although, there is significant increase in CO_2 viscosities, the amount of entrainers added is extremely high [6].

A group of researchers at the University of Wyoming attempted at in situ polymerisation of CO_2-soluble alkene monomers, including ethylene, octene, and decene [23]. They found that polymerised monomers could be miscible in CO_2 at the tested conditions (306 K and 17.9 MPa). However, the polymers did not form stable solutions and produced solid precipitates over time. As a result, viscosity enhancement was not observed. In an attempt to obtain a very high molecular weight polymeric thickener for CO_2, researchers at Chevron [24–26] have assessed candidates that exhibited Hildebrand solubility parameters of less than 7 $(cal/cc)^{0.5}$ that is closer to the CO_2 solubility parameter at reservoir conditions (327 K and 17.2 MPa), which is approximately 6 $(cal/cc)^{0.5}$ [27]. Furthermore, they found beneficial, if the polymer candidates contained electron donor atoms such as oxygen, nitrogen, and sulphur that are capable of interacting favourably with the carbon atom (i.e. an electron acceptor) within the CO_2 molecules. Electron donor functional groups used in this study included ethers, sulphones, siloxanes, thioethers, silyethers, esters, carbonyls, dialkylamides, and tertiary amines. The researchers concluded that such functional groups associated within polymers would enhance the solubility of the polymers to some extent through specific interaction with CO_2. However, full dissolution of such polymers in CO_2 could not be attained without the addition of toluene as a co-solvent [24].

High molecular weight polydimethylsiloxane (PDMS, Mw: 135,000 $g.mol^{-1}$) (**Figure 1**) was initially tested by Heller et al. [11] for solubility in CO_2. They found that 0.03 wt% of PDMS could dissolve in CO_2 at 298 K and 18.9 MPa. However, at this dilute concentration, PDMS did not appreciably enhance CO_2 viscosity. Furthermore, others attempted to increase the PDMS concentrations in CO_2; however, the solubility of PDMS in CO_2 could not be attained without substantial addition of toluene as a co-solvent. Therefore, it was determined that very high molecular weight PDMS (Mw: 197,000 $g.mol^{-1}$ and 7.3 $(cal/cc)^{0.5}$) could effectively thicken CO_2 only if a toluene co-solvent (10–20%) was added into the solution [26]. For example, addition of 20 wt% toluene co-solvent enabled up to 4 wt% of PDMS to be dissolved in CO_2, resulting in a 30-fold of CO_2 viscosity enhancement [26].

This viscosity enhancement was compared only with pure CO_2 viscosity, but it was not compared against toluene/CO_2 viscosity, because it was expected that toluene addition into CO_2 may not contribute directly to the CO_2 viscosity enhancement and it only improves the solubility of polymer in CO_2. However, their core flooding experiment results showed that CO_2/toluene flood gives higher oil recovery than pure CO_2 flood. This attributes to that toluene is a strong solvent, which causes a higher oil swelling and oil viscosity reduction. It was also found the viscous solution (20 wt% toluene, 4 wt% of PDMS, and 76 wt% of CO_2) can significantly reduce gas mobility, increase oil recovery (10–20%), and delay the breakthrough in porous media. In another study a group of researchers from Texas A&M University added an organic co-solvent (e.g. toluene) into CO_2-philic polymeric thickeners (PDMS and polyvinyl acetate (PVAc)) during core flooding experiments in order to enhance the solubility in CO_2 [28, 29]. They prepared solutions of PDMS (5 wt% with a Mw of 260,000 $g.mol^{-1}$) and PVAc (5 wt% with a Mw of 170,000 $g.mol^{-1}$) with a range of toluene concentrations from 10 to 20 wt% added as a co-solvent. In addition, PVEE was used at the concentration of 0.8 wt% in CO_2 without the addition of a co-solvent, as this polymer has the ability to dissolve in CO_2 without a co-solvent [29]. Their results proved that PDMS and PVAc with the addition of toluene could improve the gas mobility, accelerate the oil recovery (5–10% with PDMS and 4–9% with PVAc), and delay CO_2 breakthrough. These results were consistent with the finding of Chevron researchers that 4 wt% PDMS was soluble and could thicken CO_2 in the presence of a co-solvent. In other words, both groups found that PDMS and PVAc are both to be CO_2-philic and effective thickeners with the use of substantial amounts of a co-solvent. PVEE (Mw: 3800 $g.mol^{-1}$) at low concentration (0.8 wt%) did not show any increase in viscosity or improvement in CO_2 mobility and oil recovery [29]. This means that the PVEE may not be able to increase the CO_2 viscosity at this concentration which further confirms that the thickening level reported by Zhang et.al is higher than expected.

A group of researchers at the New Mexico Petroleum Recovery Research Center (PRRC) proposed another route towards high-performance thickeners by introducing self-interacting functional groups at each end of the polymer chains [14]. These polymers with terminal ionic groups (linear, difunctional, and telechelic ionomers) were thought to be effective thickeners in non-polar solvents as the ionic groups in each chain could aggregate into multiple ion pairs causing interaction between separate chains. They incorporated sulphonate groups onto the ends of polyisobutylene, and although this polymer is soluble in CO_2 at a concentration of 0.4 wt%, the sulphonated ionomer was only partially dissolved in CO_2, which is a weak solvent for this ionic group. In a recent publication, O'Brien et.al [30] used a route similar to the approach proposed by the PRRC group. The difference between the two studies is that the PRRC group used functional groups in the polymer chain that experiences intermolecular interactions with CO_2, while O'Brien proposes using functional groups that form self-assembly and intermolecular interactions via hydrogen bonding (donor-acceptor), π-π stacking, and dipole-dipole interactions. Therefore, the strategy used by the O'Brien's group could be used to explore oligomeric molecules. So, they synthesised a series of aromatic-amide terminated PDMS oligomers (i.e. low molecular weight polymers) to maximise the entropic characteristics for oligomeric species interacting with CO_2 by choosing a solute with high free volume and flexible chains. In addition, the aromatic moieties promote the formation of supramolecular structures among the low molecular weight oligomers. Amide and aromatic amide further enhance this interaction and induce self-assembly through strong hydrogen bond donor-acceptor interactions. The researchers found that amide-terminated-PDMS oligomers with simple aromatic groups through the incorporation of electron-deficient aromatic groups onto these amides (i.e. 4-nitrobenzamide, biphenyl-4-carboxamide, and anthraquinone-2-carboxamide) did not show any significant impact on CO_2 viscosity at a concentration of 1 wt% due to the poor interaction of functional groups in these compounds with CO_2. However, they achieved promising results with anthraquinone-2-carboxamide (AQCA) as an end group that was found to be a gelator of hexane at a concentration of 15 wt%, and at concentrations ranging from 5 to 10 wt% in hexane, a significant increase in viscosity was observed. However, this behaviour did not extend to other similar compounds based on either biphenyl-4-carboxamide or 4-nitrobenzamide end groups. Therefore, they attempted to improve the intermolecular interaction with the AQCA end group by utilising branched anthraquinone amides, where the number of AQCA end groups per molecule can be increased. It was found that branched anthraquinone amides were insoluble in CO_2 at concentrations of 1 wt%. However, it was soluble in CO_2 when hexane as co-solvent was added. Hence, this branched compound can be a useful thickener in the presence of substantial amount of hexane as co-solvent. For example, at a temperature and pressure of 348 K and 34.5 MPa, respectively, a transparent solution composed of 13.3% branched anthraquinone-amide functionalised oligomers, 26.7% hexane, and 60% CO_2 was found to have a viscosity three times greater than that of a CO_2/hexane mixture without a thickener. Given the low-viscosity enhancement (threefold) and high concentration of this compound and the co-solvent required, this compound was not considered to be economical and practical for CO_2 flooding. These compounds that have considerable intermolecular interactions with CO_2 are further discussed in the next section of this chapter. Overall, all studies found that high/low molecular weight PDMS polymers were more CO_2-philic than hydrocarbon-based polymers [31], although they were not capable of viscosifying CO_2 without the use of substantial amounts of a co-solvent. However, the high cost of PDMS polymer/oligomer and high concentration of co-solvent required make the field application for this polymer impractical [2, 6, 7].

DeSimone's research group [32] was arguably among the first to report on a high molecular weight polymer-based CO_2 thickener capable of increasing the viscosity without the need of a co-solvent. They found that 3.4 wt/vol% of either poly(1,1-dihydroperfluorooctyl acrylate) (PFOA) (**Figure 1**) or polyfluoroacylate (PFA, Mw: 1,400,000 $g.mol^{-1}$) could be dissolved in CO_2 increasing the viscosity of CO_2 by a factor of 2.5 at a pressure of 31 MPa and temperature of 323 K. **Figure 2** shows the increase in CO_2 viscosity resulted from dissolving 3.7 and 6.7 wt/v% of PFOA at 323 K. This is the first example of high Mw polymers that can be dissolved in CO_2 and significantly thicken CO_2 in the absence of a co-solvent. To date, PFOA is still recognised as the most soluble polymer in CO_2 and among the most effective thickeners of CO_2. Unfortunately, PFOA is a fluoropolymer type, which makes it relatively expensive. Furthermore, fluorinated polymers possess environmental concerns as they are suspected as carcinogen [33]. Therefore, if the cost and environmental constraints are considered, PFOA is not practical for field application in CO_2 flooding [2].

To limit these negative aspects of fluorinated polymers and potentially make them viable, Enick and Beckman and other co-workers at the University of Pittsburgh have tried to reduce the amount of fluorinated polymers needed without affecting its performance [4, 34, 35]. They prepared a copolymer based on a perfluoropolyacrylate and a functional group, which engages strongly in intermolecular interactions, in order to promote an increase in CO_2 viscosity. This copolymer is composed of 71–79 mol% of fluoroacrylate monomer (1,1,2,2-ttatrahydro heptaecfluorodecylacrylate) and 21–29 mol% of styrene group (polyfluoroacrylate styrene or polyFAST) (**Figure 1**). The fluoroacrylate monomer is highly CO_2-philic and facilitates polyFAST solubility in CO_2. The associating styrene

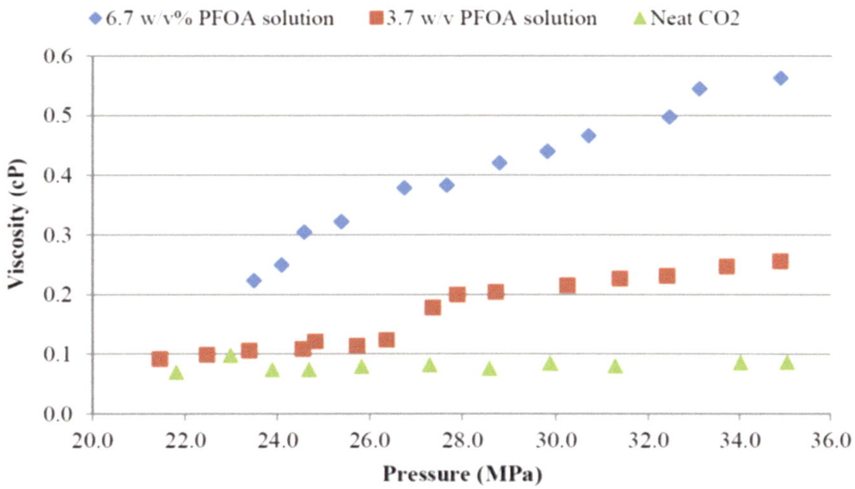

Figure 2.
The viscosity enhancements of PFOA in CO_2 at different pressures and concentrations and temperature of 323 K [32].

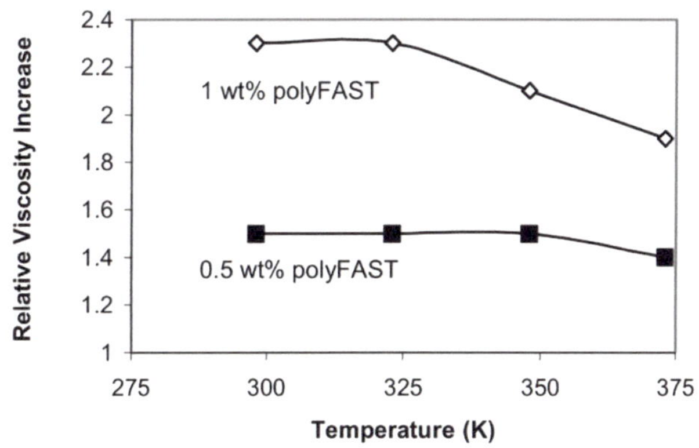

Figure 3.
The effect of temperature on the relative viscosity of polyFAST in CO_2 solution at 34 MPa [35].

group is a mildly CO_2 phobic monomer that promotes intermolecular interactions and improves viscosity enhancement through supramolecular interactions. This copolymer was found to be soluble in CO_2 at pressure and temperature conditions close to those used in CO_2-EOR [34]. However, the solubility was found to decrease with an increase in the styrene content [4]. For instance, the cloud point pressure of 1 wt% of 29 mol% styrene-71 mol% fluoroacrylate copolymer and 35 mol% styrene-71 mol% fluoroacrylate copolymer in CO_2 at 297 K is 12 MPa and 16.2 MPa, respectively. Furthermore, it was also found to significantly increase CO_2 viscosity at dilute concentrations of polyFAST. As it can be seen in **Figure 3**, 0.5 and 1 wt% of polyFAST in CO_2 at 298 K are able to increase the viscosity of CO_2 by 1.5- and 2.3-fold, respectively [34]. However, polyFAST is the most effective polymeric thickener for CO_2 at dilute concentration in the absence of a co-solvent. Comparing to PFOA results, this copolymer was successful to reduce concentration by 73% to achieve the same viscosity improvement (2.3-fold) at 323 K and 34 MPa. However, it was not practical to be used for CO_2-EOR application due to the cost of the copolymer (roughly $132 per kg) and lack of its availability in large quantities [2]. In addition, this copolymer contains a large amount of fluorine, which is environmentally and biologically persistent [29].

Another promising strategy to obtain effective CO_2 thickeners was introduced to avoid the aforementioned environmental and economic concerns associated with fluorous and silicone-based polymers. Several researchers have focused on the synthesis and design of non-fluorinated oligomers and polymers. Tapriyal et al. [36] found that PVAc is the second most CO_2 soluble polymer among non-fluorous polymers with PDMS being the most soluble. However, the dissolution of high Mw PVAc in CO_2 requires a very high pressure. In addition, no measureable viscosity increase was observed with 1–2 wt% of PVAc (Mw: 11000 $g.mol^{-1}$) in CO_2 at 298 K and 64 MPa. Furthermore, Enick and co-workers [36] designed a non-fluorinated version of PFOA in the hope of finding a thickener candidate that is cheap, environmentally friendly, and capable of increasing CO_2 viscosity at low concentration. Therefore, they developed new copolymers based on an oxygenated hydrocarbon polymer (making it CO_2-philic) and a self-interacting group (or CO_2-phobic) to enhance viscosity. Some of the attractive oxygenated hydrocarbon monomers include vinyl acetate, alkyl vinyl ether, carbonyl, and sugar acetate functional groups [37–43]. Oxygenated hydrocarbon monomers containing functional groups with one or more oxygen atoms can induce thermodynamic interactions with CO_2. These oxygen atoms are electropositive, while the carbon atoms in CO_2 are electronegative, which facilities Lewis acid-base interactions. In addition, the hydrogen bond in the polymer backbone or side chain having increased the positive charge (H¨O) acts as Lewis acids towards electron the oxygen atoms in CO_2 [41]. As mentioned above, PVAc is among the most CO_2-philic high MW oxygenated hydrocarbon polymers [43]. Therefore, Enick and co-workers replaced the fluoroacrylates in polyFAST with vinyl acetate monomers in order to reduce the cost of the polymer. They designed non-fluorous copolymers for CO_2 solubility, while the styrene group was replaced with a benzoyl group for intermolecular association. This approach simplifies the copolymer synthesis as styrene cannot be polymerised with the vinyl acetate monomer due to the large reactivity ratio difference [36]. They synthesised a 5% benzoyl-95% vinyl acetate copolymer or polyBOVA (Mw: 7840 $g.mol^{-1}$). A modest increase in CO_2 viscosity of 40–80% at a concentration of a 1 and 2 wt% was observed; however, high pressure was required (64 MPa) to attain the dissolution of this copolymer in CO_2 at 298 K.

2.2 Small-molecule self-associating thickeners

An alternate strategy to increase the viscosity of CO_2 is to employ self-interacting low molecular weight compounds as thickeners (**Table 1**). In order to differentiate between co-solvents and this class of thickeners, these compounds interact with each other and the CO_2 resulting in a self-assembled structure that contains a solvent-philic group and a solvent-phobic segment, while a co-solvent is nonassociating and consists of solely a solvent-philic segment [7]. These small molecules do not have the required molecular weight to substantially increase the gas viscosity; however, these compounds self-associate forming a supramolecular network that enhances the CO_2 viscosity [2]. In general, these compounds contain functional groups with both CO_2-philic segments that promote dissolution, and the CO_2-phobic moieties induce intermolecular association [44]. Therefore, the various associations between neighbouring molecules within the CO_2 matrix lead to viscosity enhancement [2]. Furthermore, the self-assembly of these molecules in solution can be characterised via a dramatic viscosity change or small-angle neutron scattering (SANS), FT-IR, circular dichroism, X-ray diffraction, electron microscopy, or differential scanning calorimetry [45–53]. To date, small associative molecules have yielded little success in thickening CO_2, because CO_2 is a poor solvent for these ionic and polar associating groups [8]. The following section describes the different types of small molecules used to thicken CO_2.

Small-molecule compound	Concentration in CO_2	Co-solvent	Soluble in CO_2 observations	CO_2 viscosity increased at 298–313 K	Reference
Trialkyltin fluoride	0.13 wt%	Pentane (39 wt%)	Insoluble	No viscosity increased	[56]
Semi-fluorinated trialkyltin fluorides and fluorinated telechelic ionomers	2–4 wt%	No co-solvent	Soluble	2–3-fold	[55]
Hydroxyaluminum disoaps	—	—	Insoluble	No	[62]
Fluorinated hydroxyaluminum disoaps	—	—	Partially soluble	No	[2]
Metallic stearate powders	—	—	Insoluble	No	[62, 63]
Semi-fluorinated powders	5–20 wt%	No co-solvent	Soluble	Gel solution	[64]
Hydroxystearic acid	3 wt%	Ethanol (15 wt%)	Soluble	100-fold (gel solution)	[65]
Fluorinated bis-ureas	5 wt%	No co-solvent	Soluble	3–5-fold	[66]
Non-fluorous bis-ureas	1 wt%	No co-solvent	Soluble (opaque solution)	No viscosity measured	[67]
Sodium pentadecfluoro-5-dodecyl sulphate (NaF_7H_4)	4.4 wt% NaF_7H_4 10 mol of water/mol surfactant	No co-solvent	Soluble	2-fold	[70]
Nickel bis-nonofluoropentane sulphosuccinate (Ni-diHCF_4)	6 wt% Ni-diHCF_4 10 mol of water/mol of surfactant	No co-solvent	Soluble	1.5-fold	[49]
Branched benzene trisurea	0.5–2 wt%	Hexanes (18–48.4 wt%)	Soluble	3–300-fold	[75]

Table 1.
Outline of the solubility in CO_2 and CO_2 thickening capability of small-molecule compound.

2.2.1 Trialkyltin fluorides and semi-fluorinated trialkyltin fluorides

Heller and co-workers studied a series of trialkyltin fluoride compounds as light alkane and CO_2 thickeners [12, 13, 44]. These compounds show a moderate increase in CO_2 viscosity via the formation of intermolecular associations between the tin and fluorine atoms in the solution. **Figure 4** shows the association of tributyltin fluoride molecules. Trialkyltin fluoride forms a long linear transient polymeric chain through intermolecular association between the tin atom and the fluorine

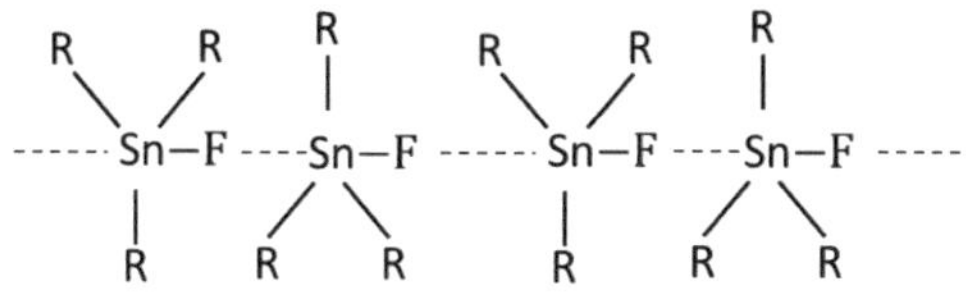

Figure 4.
Association mechanism of tributyltin fluoride [54].

atom of neighbouring molecules. In fact, the tin atom is slightly electropositive which interacts with the electronegative fluorine atom to form an intermolecular Sn-F association, as can be seen in **Figure 4**, while the hydrocarbon arms branching from the tin atom enhance the free volume which facilitates the solubility in CO_2 [55]. Apparently, these molecular structures form linear and associating structures in which the alkyl arms stabilise the aggregation, while the tin atoms in each molecule associate with the fluorine atoms in adjacent neighbour molecules [55]. Although there was some success with tributyltin fluoride or other trialkyltin fluorides in thickening light alkane components, these compounds were insoluble in CO_2 and ineffective as thickeners, even with the addition of pentane as a co-solvent [56, 57]. Later on, Shi et al. [55] synthesised semi-fluorinated trialkyltin fluorides and fluorinated telechelic ionomers to prepare a solution containing both CO_2-philic fluorinated groups to enhance the solubility and CO_2-phobic associating group to promote intramolecular association for viscosity enhancement. Both ionomers were soluble in CO_2 at 2–4 wt% without requiring the addition of a co-solvent. Their results indicated that both ionomers were capable of increasing the viscosity of CO_2 by 2–3-fold over a concentration range of 2–4 wt%. For example, at 4 wt% of tri(2-perfluorobutyl ethyl) tin fluoride in CO_2, the viscosity increased three times at 298 K and 16.5 MPa. This viscosity increase was found to be much less than expected because the side chain fluorine atoms on the Sn-F associations were disrupted. This is attributed to the fluorine atom at the end alky arms competing with the fluorine atom attached to the tin atom caused by the electronegativity differences between these chain-end fluorines and those adjacent to the tin. Hence, the disruption of the fluorinated alkyl chains is responsible for the viscosity increase [55]. Overall, given the necessary high concentrations of the ionomers required and their high costs, these fluorinate oligomers are not considered viable thickeners for field applications [2, 6, 55].

2.2.2 Fluorinated and non-fluorous hydroxyaluminum disoaps

Hydroxyaluminum disoaps were developed to thicken gasoline which was used to make napalm, the infamous weapon type used in World War II [58–60]. These molecules are an aluminium-based soap with two carboxylic acid groups linked to the aluminium atom [61]. A small amount of hydroxyaluminum disoap added to low-viscosity gasoline transforms it to a thick and extremely viscous fluid referred as napalm. In an analogous manner, these compounds were studied to determine their solubility in CO_2 and quantify their ability to thicken CO_2. Enick and co-workers synthesised a series of hydroxyaluminum disoaps [62]. Unfortunately, none of the hydroxyaluminum disoaps were soluble in CO_2. Similar to the results with trialkyl tin compounds summarised above, unpublished results by Enick showed that the solubility of some of these compounds in CO_2 could be enhanced either by fluorinating the alkyl arms or using highly branched alkyl chains [2]. However, this trial has not been successful in fully dissolving the hydroxyaluminum disoaps in CO_2 [2]. Another attempt to thicken CO_2 was done by heating a mixture of CO_2 and metallic stearate powders [63]. Metal stearates are salts that are produced from the reaction of stearic acid and metal oxide, which are dissolved in hydrocarbon-based oils usually with the assistance of heat to break up strong intermolecular forces. The viscosity of the hydrocarbon-based oils is enhanced when the solution cools down. This same approach was attempted with CO_2; however, this was unsuccessful as they are insoluble even with the assistance of heat.

2.2.3 Semi-fluorinated alkanes

Iezzi and co-workers [64] made an early attempt to thickened CO_2 by using semi-fluorinated alkanes. They designed a series of linear diblock alkane compounds ($F(CF_2)_n (CH_2)_m H$), which contained two immiscible segments forced to interact via a covalent carbon-carbon bond. It is found that this compound can gel organic liquid (e.g. decane and octane) through the formation of microfibrillar network if the solution is heated and then left to cool down. After the solution (CO_2 and semi-fluorinated alkane) cools, the semi-fluorinated compounds form a covalent cross-link between molecules and high-porosity and micro-fibrillar networks that can gel the dense CO_2. The fluorinated segments stack with other adjacent fluorinated segments (analogous to hydrocarbon segments) to form the fibre network [7, 56]. However, the gel solution is not suitable for gas mobility control due to its phase behaviour where the viscous solution could not flow through a porous medium and retained at the surface of the rock. This solution may be applicable for conformance control to block fractures or high permeable zones.

2.2.4 Hydroxystearic acid

Heller and co-workers [65] proposed a small organic compound, known as 12-hydroxystearic acid (HAS). This compound ($H_3C(CH_2)_5 CHOH(CH_2)_{10}COOH$) had previously been used to gel hydrocarbon and chlorinated solvents. However, the essential assessment results indicated that the HAS was insoluble in CO_2 unless a significant amount of a co-solvent (i.e. ethanol) was added. For instance, the addition of 15 wt% of ethanol co-solvent enabled a solubility of up to 3 wt% of HAS in CO_2 resulting in a nearly 100-fold increase in viscosity in the temperature range of 300–307 K. As the temperatures decrease, the solution exhibits a slight viscosity increase. In addition, microfibres in the gel solution form an opaque solution that would probably impede the fluid flow in the reservoir formation.

2.2.5 Fluorinated and non-fluorinated bis-ureas

A group of researchers at Yale University and the University of Pittsburgh developed a series of small-molecule compounds containing either one or two urea groups [66]. The urea groups in these compounds induce self-assembly interactions via a hydrogen bond; thereby, these interactions form macromolecular associations that can enhance the viscosity of the CO_2-rich solution. Out of the 12 compounds tested, 4 fluorinated bis-urea compounds were highly soluble in CO_2 without needing heat and capable of improving the CO_2 viscosity by 3–5-fold at 5 wt% of bis-urea at 298 K and 31 MPa. In the hope of obtaining a non-fluorous bis-urea, Paik et al. [67] attempted to incorporate the CO_2-philic groups (hydrocarbon, carbonyl, and ether groups) into the molecular structure of the bis-urea as illustrated in **Figure 5**. However, their results revealed that after forming a transparent solution, microfibres began to form slowly due to the molecules undergoing self-assembly and precipitating out of solution. Therefore, these compounds cannot be considered for EOR applications.

2.2.6 Surfactants with twin and divalent metal cations

Eastoe and co-workers designed semi-fluorinated surfactants based on a previous study that used aerosol-OT (AOT)-based water-in-oil microemulsions in cyclohexane solvent [49, 68, 69]. The molecular structure of these surfactants is illustrated in **Figure 6**. These surfactants are soluble in CO_2 and form rodlike micelles that enhance CO_2 viscosity with the addition of a small amount of water. The purpose of adding water into the solution is to form a stable microemulsion in the presence of AOT surfactant and to promote an aggregate shape that changes the self-association of the surfactant from a spheroid (non-viscous) configuration to a rod shape (viscous) that significantly enhances the viscosity of the solution. Two di-chain perfluorinated sulphosuccinate surfactants (nickel bis-nonofluoropentane sulphosuccinate (Ni-diHCF_4) and sodium pentadecfluoro-5-dodecyl sulphate (NaF_7H_4)) yielded the greatest viscosity increase among the compounds tested. These surfactants have been modified in such a way that can form rod-like micelles to promote solubility in CO_2 and viscosity enhancement. This was achieved by the exchange of Na^+ ions with Ni^{2+} or Co^{2+} to drive a sphere-to-rod

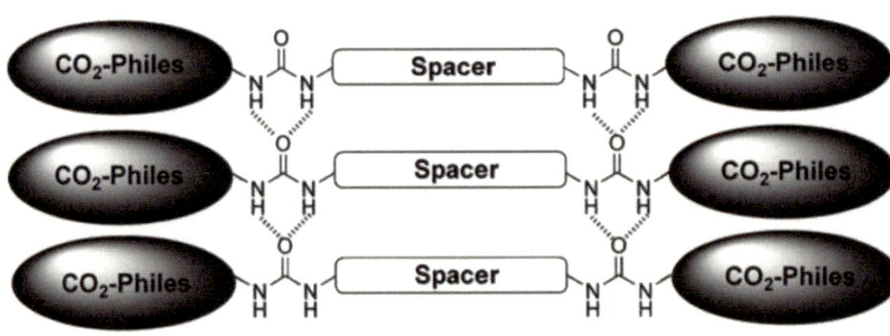

Figure 5.
Molecular structure of non-fluorous bis-ureas [67].

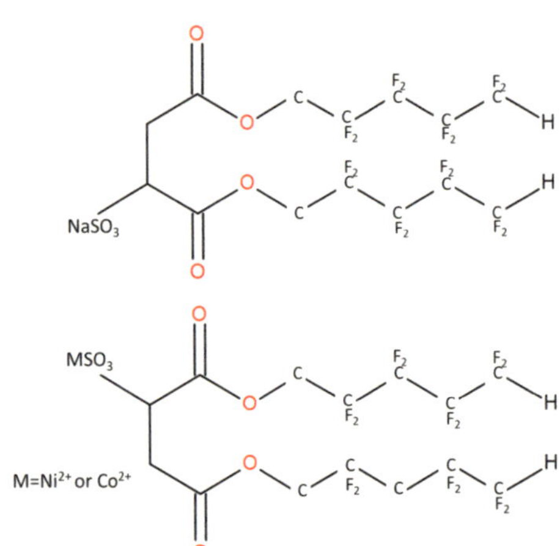

Figure 6.
Molecular structure of a fluorinated twin-tailed surfactant as a CO_2 thickener [49].

transition as shown in **Figure 6** [49, 70]. Furthermore, a di-chain perfluorinated AOT analogue is known to stabilise microemulsions of water in CO_2 [71–74]. A high-pressure small-angle neutron scattering (SANS) confirmed the solubility of both surfactant in CO_2 and formation of rodlike micelles. At 298 K and 40 MPa, both surfactants (0.05 mol dm^{-3}) with 10–12.5 moles of water per mole of surfactant achieved a transparent solution in CO_2 [49, 70]. At 298 K, 35 Mpa, 6 wt% of Ni-diHCF_4, and 10 moles of water per mole of surfactant added into CO_2, viscosity enhancements of up to 1.5-fold have resulted [49], and 4.4 wt% of NaF_7H_4 with 12.5 moles of water per mole of surfactant caused a 2-fold increase in viscosity at 313 K and 40 MPa [70]. However, these surfactants required a very high pressure to attain a single phase, and high concentrations of 5–7 wt% were necessary to achieve a significant viscosity increase. Therefore, these thickeners would not be suitable for field applications as both need a relatively high concentration of these expansive surfactants.

2.2.7 Cyclic and aromatic amide and urea based

Most of the successful associating small-molecule compounds as CO_2 thickeners that have been described above are fluorinated or semi-fluorinated materials. These fluorinated materials are both expensive and environmentally persistent due to the fluorine content and high concentrations (3–5 wt%) required for use as a CO_2-EOR thickener [75]. Therefore, in a recent publication, Doherty et al. [75] synthesised and examined a series of cyclic and aromatic amide and urea compounds as non-fluorous small-molecule thickeners for dense CO_2 and organic liquids. They designed the molecular structure of the compounds as shown in **Figure** 7. These compounds contain cyclic or aromatic core molecules (e.g. cyclohexane or benzene) which are mildly CO_2-phobic to promote intermolecular interactions. These core ring groups are combined with associating or linking groups (labelled as 'X') which are typically either amide, urea, or ester groups to establish the intermolecular interaction for viscosity enhancement. In addition, these linking groups also aid the connection of CO_2-philic segments (siloxane or heavily acetylated) to cyclic or aromatic core molecules to improve dissolution in CO_2. It has been found that after heating and cooling the mixture, these compounds were capable of thickening organic liquids such as hexane and toluene. Researchers have found branched benzene trisurea (propyltris(trimethylsiloxy)silane-functionalised benzene trisurea and trisurea compounds functionalised with varying proportions of propyltris(trimethylsiloxy)silane and propyl poly(dimethylsiloxane)-butyl groups), which are soluble in dense CO_2. These compounds are capable of thickening CO_2 (3–300-fold) at remarkably low concentrations (0.5–2 wt%) in the presence of hexane as a co-solvent at high concentrations (18–48 wt%) [2, 8]. A 300-fold viscosity increase is too large and definitely not suitable for EOR applications. The high concentration of the required co-solvent at low concentration of the additive severely limits the applicability of this approach due to the high manufacturing costs and environmental concerns (**Table 1**).

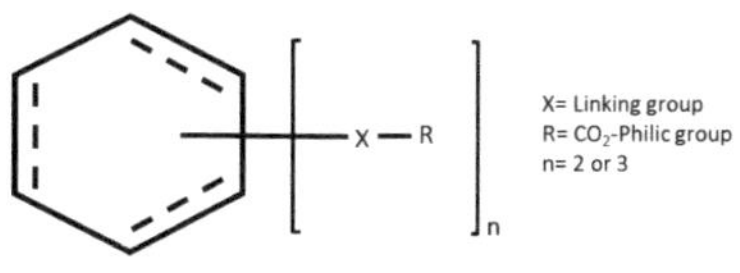

Figure 7.
General molecular structure of small-molecule cyclic amide and urea based [75].

3. Hydrocarbon gas thickeners

3.1 Polymeric thickeners

As discussed so far, most of the research regarding gas thickening agents have focused solely on CO_2 because it is the most common injected fluid for MGI projects in the United States, Canada, and elsewhere [76, 77]. In addition, CO_2 is a slightly more powerful polymer solvent than short-chain alkane gases. The structural symmetry of CO_2 results in a substantial quadruple moment (Q_i) at low pressure and temperature, which can magnify the quadrupole interaction by scale inversely with the molar volume to the 5/6 power ($Q_i^* = Q_i . V_i^{-5/6}$) [78]. Despite these characteristics, CO_2 is a weak solvent when compared to most organic solvents. However, there have been a few attempts at identifying polymeric thickeners for pure light hydrocarbon gases [79–81]. In the late 1960s, several patents reported initial attempts at thickening light alkane gases. Henderson et al. [81] made the first attempt to thicken a hydrocarbon by using three polymers including poly methyl laurylate, polybutadiene, and poly(alkyl styrene). These polymers (at a concentration of 0.25 vol%) are capable of improving the viscosity of light hydrocarbon gases by about 0.1%. Subsequently, Durben and co-workers examined polyisobutylene polymer (PIB, Mw: 130,000 $g.mol^{-1}$) in a rich condensate mixture containing 75 vol% propane and 25 vol% heptane. They claimed to achieve a 2–5-fold viscosity increase at a concentration of 0.25 wt% of PIB [80]. However, none of the patented work reported the details of the method used to measure the viscosity of the solutions examined.

Subsequent attempts by Heller et al. to identify polymeric thickeners for LPG and CO_2 [9] found that various poly-α-olefin polymers (PAO) based on n-pentene, n-hexene, and n-decene could be used. These polymers were found to be quite soluble in n-butane at a temperature of 298 K and pressure of 8.2 MPa; however, their solubility in CO_2 was much more limited at a temperature of 305 K and pressure of 17.2 MPa. The addition of these polymers at concentrations ranging from 1 to 2.2 wt% to n-butane enhanced the viscosity by fivefold (**Table 2**). In a recent publication, Dhuwe et al. assessed the solubility and viscosity-enhancing property of high and ultrahigh molecular weight polymers in NGL (i.e. a mixture of ethane, propane, and butane) [82, 83]. Polymers evaluated included ultrahigh molecular weight drag-reducing agent (DRA) poly-α-olefin (Mw: 20,000,000 $g.mol^{-1}$), high molecular weight PDMS (Mw: 980,000 $g.mol^{-1}$), and PIB (Mw: 130,000 $g.mol^{-1}$). Ultrahigh molecular weight DRA poly-α-olefin is commonly used in oil pipelines to supress the energy dissipation near the pipe wall that results from the turbulent flow at high flow rates. This polymer does not change the fluid properties (e.g. viscosity) at the dilute concentrations used for this application. Dhuwe et al. [83] found it to be sufficiently soluble in NGL if a significant amount of hexane is added as a co-solvent. For example, at 0.5 wt% of DRA polymer and 24.5 wt% hexane in propane or butane, the cloud point pressures at temperatures of 333 K were found to be equal to 3.07 and 0.77 MPa, respectively. However, it requires very high pressure to attain solubility in ethane (46.95 MPa) at the same concentrations. At 0.5 wt% of DRA polymer and 24.5 wt% of hexane, the viscosity of ethane and propane could be improved by 3–9-fold, while 23–30-fold enhancement was obtained in butane (**Table 2**). The reason for the greater increase in the viscosity of butane is explained by the larger butane solubility compared to the solubility of propane and ethane that aids the expansion of the polymer backbone (i.e. coil) that swells the DRA polymer [82, 83].

Furthermore, they have also tested the solubility of high molecular weight PIB and PDMS in NGL components. PIB was found to be insoluble in ethane, propane, and butane at temperatures ranging from 298 to 353 K and high pressure, while PDMS was soluble in all NGL constituents without the aid of a co-solvent [82, 83].

DOI: http://dx.doi.org/10.5772/intechopen.88083

Small-molecule compound	Concentration in NGL components (wt%)	Co-solvent	Soluble in NGL observations			Relative viscosity at 298–333 K			Ref.
			C_2H_6	C_3H_8	C_4H_{10}	C_2H_6	C_3H_8	C_4H_{10}	
Poly(1-pentene), poly(1-hexene), and poly(1-decene)	1–2.2	No	—	—	Yes	—	—	5-fold	[9]
Drag reducing agent (DRA) poly-α-olefin	0.5	24.5 wt% of hexane	Yes	Yes	Yes	3-fold	9-fold	23–30-fold	[82, 83]
Poly(dimethyl siloxane)	2	No	Yes	Yes	Yes	1.2-fold	2-fold	4-fold	[82, 83]
Tri-n-butyl tin fluoride	1	No	Yes	Yes	Yes	1.5–4-fold	40-fold	60-fold	[82, 88]
Hydroxyaluminum di-2-ethylhexanoate (HAD2EH)	0.5–1	No	No	Yes	Yes	No	10-fold	14-fold	[82, 88, 89]
Cross-linked phosphate esters (HGA 70-C6) + HGA 65	0.25–1	No	Yes	Yes	Yes	1.45-fold	2.6-fold	3.5-fold	[88]

Table 2.
Outline of the solubility in NGL compounds and NGL thickening capability of polymers and small-molecule compounds.

In propane and butane, PDMS was soluble at pressures close to the vapour pressures of propane and butane, while in ethane, it required high pressure (much greater than its vapour pressure) to attain solubility. For example, at 333 K and 2 wt% of PDMS in ethane, propane, and butane, the cloud point pressures were obtained to be equal to 18, 2.56, and 0.92 MPa, respectively. Furthermore, they found PDMS to be an effective thickener in propane and butane but an ineffective thickener in ethane. For example, at 333 K and at a concentration of 2 wt%, PDMS achieved viscosity increases of 1.2-fold in ethane, 2-fold in propane, and 4-fold in butane. It was also found to be a better thickener at high pressure (62 MPa). Overall, high molecular weight PDMS is not a viable thickener of NGL for EOR applications [83].

In comparison to the results obtained by Heller et al. for 2.2 wt% of PAO (poly(1-pentene), poly (1-hexene), and P-1-D) in butane at 298 K, DRA increases butane viscosity substantially at even lower concentrations (0.5 wt%), because of the extremely high molecular weight of DRA, regardless the relatively high concentration of hexane added to the system. Although PDMS has a higher molecular weight (Mw: 980,000 g.mol^{-1}) than PAO, PDMS offers a lower relative viscosity (fourfold) than the viscosity obtained by low molecular weight PAO (fivefold). These observations indicate that the increase in gas viscosity depends on several factors and not solely on the molecular weight of the additives. Additional factors that can influence the viscosity-enhancing ability of an additive include the nature of additives and the solvent, the concentration of additives, the molecular weight distribution of the additives, and the type of intermolecular interactions among the additives and the solvent [84, 85]. These chemical additives (PDMS and PAO) have different chemical structures. PAO has a carbon-carbon backbone with atactic molecular structure of mostly uniform head-to-tail connections with some head-to-head-type connections in the structure [86]. On the other hand, PDMS has silicone-oxygen backbone and more flexible molecules than P-1-D molecules. Hence, PDMS can have lower steric hindrance and greater bond angle (143° vs. 110° for C-C-C) to rotate around the Si-O bond [87]. Furthermore, the effect of molecular structure and polymer molecular weight on viscosity has been studied by Zolper et al. [87] who found that similar viscosity can be obtained for different molecular mass. For example, the viscosity of PAO at 1000 g.mol^{-1} is equivalent to the viscosity of PDMS at around 10,000 g/mol. This was attributed to the additional attractive intermolecular forces between the polymers with increasing branches, which leads PAO to having higher viscosity indices than PDMS. Therefore, the effect of PAO on butane viscosity could be attributed to the structure of the polymer. These effects are more pronounced in improving the solvent viscosity than the PAO molecular weight.

3.2 Small-molecule self-associating thickeners

Similarly to studies for CO_2, low molecular weight self-associating compounds have been studied as thickening agents of light alkane gases for gas mobility control and hydraulic fracturing purposes [54, 82, 88, 89]. Ideally, small-molecule compounds need two processes to attain dissolution and viscosity enhancement [88, 89]. The first is a high-pressure heating cycle, which disrupts the intermolecular association to enhance the dissolution. The second process is the cooling cycle to re-establish the intermolecular association necessary for viscosity enhancement. However, some small-molecule compounds do not require this two-step process to attain dissolution and viscosity enhancement in NGL [89]. Previous studies on the application of small-molecule compounds to increase the viscosity of light alkane gases are discussed in the following section.

3.2.1 Trialkyltin fluorides

Dunn and Oldfield first reported on the use of tri-n-butyl tin fluoride (TBTF) as a direct thickener of non-polar solvents including carbon tetrachloride and n-propane [89]. **Figure 8** illustrates the association mechanism of tributyltin fluoride [90], where a linear polymeric chain of penta-coordinate tin atoms are linked by fluorine atoms. TBTF is a white powder with a melting point of 544 K [88, 89]. The three butyl arms attached to the tin atom enhances the solubility of TBTF in a hydrocarbon solvent, while the intermolecular association formed among the tin and fluorine atoms induces the viscosity-enhancing effect. It has been found that TBTF is soluble in organic liquids and light alkanes under stirring for several minutes without requiring a heating and/or a cooling cycle [89]. TBTF is an effective thickener for intermediate hydrocarbon components. Dandge et al. [54] found TBTF to be capable of improving the viscosity of propane and butane. For instance, at concentrations of 0.13–0.15 wt% at 298 K, it increased the viscosity of these components by 2–10-fold at 8.3 MPa. In addition, they also found that TBTF was only partially soluble in ethane and with no measurable viscosity change [79]. Later on, Enick and co-workers confirmed the ability of TBTF to thicken propane and butane liquids at 298 K at concentrations of 0.2–5 wt% [56]. Other trialkyltin fluorides have been tested in hydrocarbon solvents. Tripropyltin fluoride (TPTF) is not soluble in propane and butane, because the propyl arms are too short to induce the dissolution of TPTF in these solvents [91]. Therefore, it confirms that the solubility of trialkyltin fluoride in n-alkane increases as the number of carbon atoms in n-alkyl arms (R) increases [54]. However, at equivalent mass concentration, TBTF in n-hexane or n-butane has shown to outperform in viscosity enhancement compared with other solvents [54]. For example, at a concentration of 10 g/L and 310 K, TBTF increases the viscosity of n-hexane by 750-fold (from 0.265 cP to 196 cP), while the tetrachloroethylene viscosity is enhanced by 380-fold (124.45 cP) [54].

A recent study has tested the solubility and viscosity enhancement ability for dilute concentrations (>1 wt%) of TBTF in ethane, propane, and butane at high pressures (38–64 MPa) and high temperatures (298–373 K) [82, 88, 89]. TBTF is soluble in propane and butane above the corresponding vapour pressure of these components, while in ethane, TBTF is soluble at much higher pressures than the ethane vapour pressure. In addition, it was observed that the relative viscosity of TBTF in NGL components increases slightly with increasing pressure at all temperatures and TBTF concentrations. Increasing the pressure does not affect the self-assembly of the supramolecular structure; it only affects the solvent strength which has a less significant effect on the solution viscosity. Furthermore, as temperature increases, the intermolecular association between the tin and fluoride

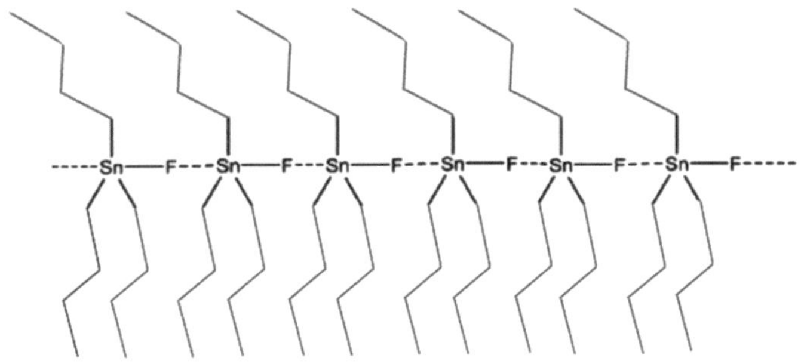

Figure 8.
Association mechanism of tributyltin fluoride [89].

molecules diminish, leading to a significant decrease in the viscosity enhancement in all light alkane components. For example, with 1 wt% concentration of TBTF in ethane at 298 K and 62 MPa, the achieved relative viscosity is 90, and it drops to 75 at 313 K. The relative viscosity significantly drops further to 20, 6, and 1.5 at 333, 353, and 373 K, respectively [82] (**Table 2**).

3.2.2 Hydroxyaluminum di-2-ethylhexanoate (HAD2EH)

A mixture of aluminium disoap and gasoline liquid is heated to high temperatures (368–373 K) to promote its dissolution by dismantling the intermolecular associations between the aluminium disoaps. Then it is cooled down to allow self-assembly of the disoap molecules, whereby the viscosity of the solution is enhanced significantly [88, 89]. Enick and co-workers [62] studied a single aluminium salt, referred to as hydroxyaluminum di-2-ethylhexanoate (HAD2EH). **Figure 9** depicted the association mechanism of HAD2EH. They found HAD2EH to exhibit a remarkable solubility in light hydrocarbon gases such as propane and butane. It is also capable of thickening these components at dilute concentrations. For example, at 293 K, HAD2EH concentrations in the range of 0.2–1 wt% were capable of increasing the viscosity of the solution by 10–100-fold as tested in a high-pressure falling-ball cylinder viscometer. However, the solution formed was hazy, due to a portion of the HAD2EH molecules forming solid fibres in both liquid propane and butane at high pressures.

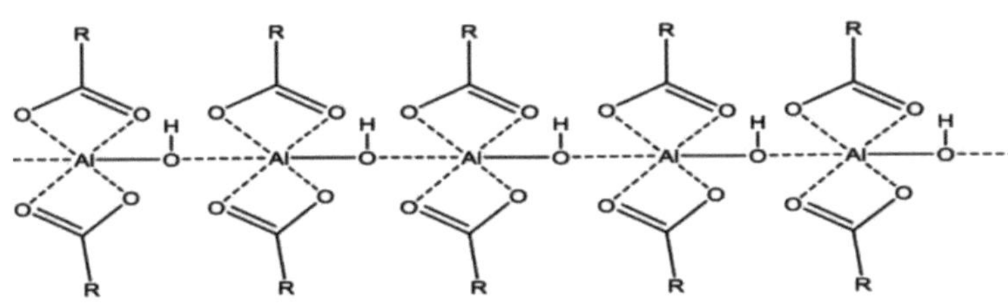

Figure 9.
Association mechanism of HAD2EH molecules [89].

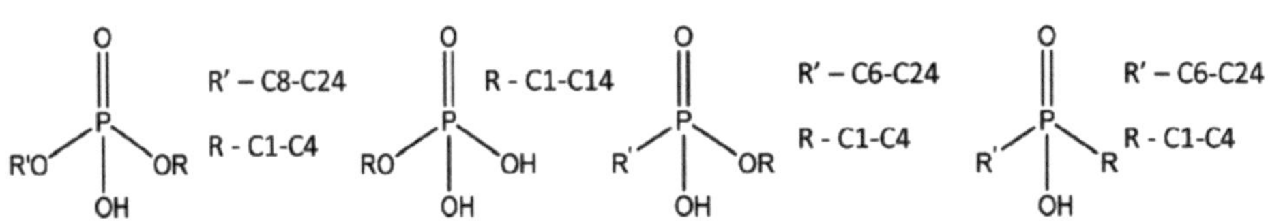

Figure 10.
Molecular structure of phosphate di-/monoester, phosphonic acid ester, and dialkyl phosphinic acid [89].

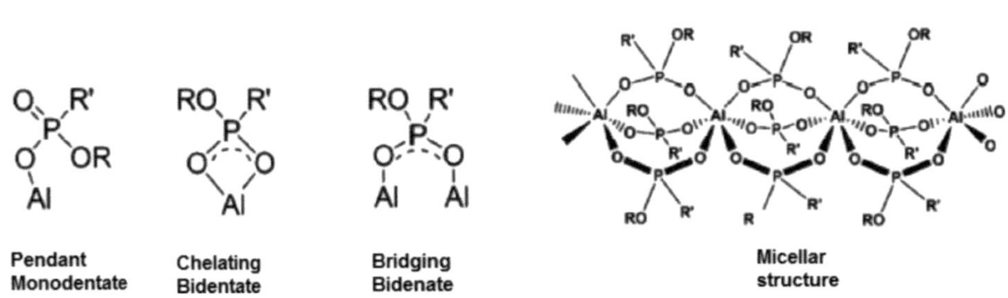

Figure 11.
Chelation mechanism and micellar structure of phosphate ester/metal ion complex [96].

Dhuwe et al. [82, 89] examined blends of HAD2EH in several NGL (i.e. ethane, propane, and butane) individually under a range of pressures (34–62 MPa) and temperatures (298–273 K). HAD2EH was soluble in propane and butane while insoluble in ethane. At a temperature of 298 K, HAD2EH was insoluble in all light alkanes. Heating to 373 K, high pressure, and stirring were required to attain dissolution, followed by cooling to a temperature above 313 K to obtain a single phase solution. If the solution is cooled down to a temperature of 298 K, HAD2EH precipitated in both propane and ethane. Accordingly, it was observed that HAD2EH was an effective thickener in butane and propane at temperatures as low as 313 K. For example, at a concentration of 0.5 wt% HAD2EH and temperatures of 333–373 K, butane viscosity increases by 15–19-fold, while propane is thickened by 2–3-fold [82, 89].

3.2.3 Cross-linked phosphate esters

Several studies report the attempts made to gel light hydrocarbon gas (LPG) using phosphorus-based esters cross-linked with polyvalent metal ions (**Figure 10**) for dry hydraulic fracturing applications [90, 92–95]. These techniques use phosphate mono-/diesters linked to alkyl tails. Typically, a hydrocarbon liquid agent solution is formed by combining two low-viscosity liquid reactants (i.e. a solution containing the phosphate ester and one containing a polyvalent metal ion cross-linking agent) together in the fluid that is being thickened. The two low-viscosity liquids quickly dissolve in the hydrocarbon fluid (e.g. light alkane) without the need of heating and/or cooling cycles. The polyvalent metal ion bound more tightly to phosphate esters than to the ligand, which leads the phosphate ester to quickly chelate with metal ion and form long micellar and a supramolecular structure as shown in **Figure 11**. If this long micellar structure remains soluble in the solvent, then it can significantly improve the solution viscosity. Rapid dissolution of the phosphate ester system in the solvent and the rapid viscosity enhancement kinetics could make these molecular structures attractive for use with NGL in EOR applications [88, 89].

There are a few studies reporting the use of oil-soluble phosphate mono-/diesters, dialkyl phosphinic acids, or alkyl phosphonic acid ester cross-linked with polyvalent metal ions including Fe^{3+}, Mg^{2+}, Al^{3+}, Zn^{2+}, and Ti^{4+} at concentrations of 0.2–2.5 wt% to increase the viscosity of hydrocarbon oils (e.g. diesel and kerosene) by 2–100-fold [93, 94, 97]. Furthermore, phosphate-based esters could be used as gel agents for CO_2 and hydrocarbon liquid mixtures [98, 99]. Lee and Dhuwe et al. [82, 88, 89] studied blends of cross-linked phosphate esters (CPE) (phosphate ester (HGA 70-C6) and cross-linker (HGA 65)) with NGL components (ethane, propane, and butane) at temperatures ranging from 298 to 373 K and pressures ranging from 13.8 to 62 MPa. They found that phosphate ester and the cross-linked solution are soluble in ethane, propane, and butane at concentrations of 0.25–1 wt% and temperatures of 298–333 K. However, the cross-linked solution was slightly hazy due to small droplets of CPE suspended in the solution. This mixture (phosphate ester + cross-linker) achieved a modest viscosity increase in ethane, while greater increases occurred in propane and butane. For example, at a concentration of 1 wt% at 333 K and 20.68 MPa, the viscosity of ethane increased by a factor of 1.45-fold, and the viscosity of propane and butane increased by 2.6- and 3.2-fold, respectively. This mixture cannot form a single phase in the NGL components due to the suspension of very small droplets of CPE in the solution and because its suitability for injection into porous media is limited [88] (**Table 2**).

4. Conclusion

Despite of over five decades of extensive work in developing and identifying viable gas thickeners for EOR applications such as fluorous and non-fluorous polymers, copolymers, and low molecular weight compounds, so far none of these materials may be considered effective thickeners for field applications [6]. In fact, the design of affordable thickeners for CO_2 or HC gases effective at low concentrations (i.e. <1 wt%) is highly challenging due the low solubility of the additives in these solvents.

The main conclusions derived from this review are as follows:

- At laboratory scale, the best CO_2 thickeners are PDMS, P-1-D, polyFAST, and PFOA for CO_2 mobility and conformance control. However, the use of these materials for field applications is not sustainable due to their elevated manufacturing costs, the large concentrations required to ensure technical success, and environmental concerns.

- There are several small-molecule compounds that have been identified as CO_2 thickeners including semi-fluorinated trialkyltin fluorides, fluorinated bis-ureas, di-chain perfluorinated sulphosuccinate surfactants, and branched benzene trisurea. In general, these compounds are capable of increasing the CO_2 viscosity between 1.5- and 300-fold at lower temperatures (298–313 K). However, most of these materials are fluorinated and can improve the viscosity only at high concentrations (2–10 wt%). In addition, the intermolecular associations displayed by these compounds are effective at low temperatures; therefore, they cannot be used at typical reservoir conditions due to the diminishing intermolecular associations at high temperature [2].

- Effective polymer thickeners for NGL components include DRA polymers, PDMS, and PAO oligomers. These polymers are much more effective for thickening butane and less effective for propane and ethane. DRA polymers are promising thickeners (3–23-fold) at dilute concentrations over temperatures of 298–333 K, while PDMS and PAO show moderate viscosity enhancements (1.2–5-fold) over the same temperatures. In addition, P-1-D seems to be a suitable thickener in an AG mixture at temperatures above 373 K.

- Small-molecule compounds such as TBTF, HAD2EH, and CPE are effective thickeners for pure light alkane components at moderate temperatures (313–333 K). Nevertheless, phase behaviour and environmental issues (fluorine content) prevent the use of these materials for EOR applications.

Author details

Nasser Mohammed Al Hinai[1,2*], Matthews Myers[3], Colin D. Wood[3] and Ali Saeedi[1]

1 Department of Petroleum Engineering, Curtin University, Perth, WA, Australia

2 Petroleum Development Oman L.L.C., Muscat, Sultanate of Oman

3 Commonwealth Scientific and Industrial Research Organization, Perth, WA, Australia

*Address all correspondence to: nasser.alhinai5@gmail.com

References

[1] Enick RM, Ammer J. A literature review of attempts to increase the viscosity of dense carbon dioxide. Website of the National Energy Technology Laboratory; 1998

[2] Lee J et al. Development of small molecule CO_2 thickeners for EOR and fracturing. In: SPE Improved Oil Recovery Symposium. Tulsa, Oklahoma: USA Society of Petroleum Engineers; 2014

[3] Dai C et al. Impairment mechanism of thickened supercritical carbon dioxide fracturing fluid in tight sandstone gas reservoirs. Fuel. 2018;**211**:60-66

[4] Huang Z et al. Enhancement of the viscosity of carbon dioxide using styrene/fluoroacrylate copolymers. Macromolecules. 2000;**33**(15):5437-5442

[5] Martini A, Ramasamy US, Len M. Review of viscosity modifier lubricant additives. Tribology Letters. 2018;**66**(2):58

[6] Enick RM et al. Mobility and conformance control for CO_2 EOR via thickeners, foams, and gels—A literature review of 40 years of research and pilot tests. In: SPE Improved Oil Recovery Symposium. Tulsa, Oklahoma, USA: Society of Petroleum Engineers; 2012

[7] Enick R, Olsen D. Mobility and conformance control for carbon dioxide enhanced oil recovery (CO_2-EOR) via thickeners, foams, and gels–A detailed literature review of 40 years of research. U.S. Department of Energy, National Energy Technology Laboratory (NETL); 2012

[8] Lee J. Small molecule associative CO_2 thickeners for improved mobility control (doctoral dissertation). Pittsburgh, USA: University of Pittsburgh; 2017

[9] Dandge D, Heller J. Polymers for mobility control in CO_2 floods. In: SPE International Symposium on Oilfield Chemistry. San Antonio: Texas Society of Petroleum Engineers; 1987

[10] Dandge D et al. Kinetics of 1-hexene polymerization. Journal of Applied Polymer Science. 1986;**32**(6):5373-5383

[11] Heller J et al. Direct thickeners for mobility control of CO_2 floods. Society of Petroleum Engineers Journal. 1985;**25**(05):679-686

[12] Heller J, Kovarik F, Taber J. Improvement of CO_2 flood performance. Annual Report 1985. 1986

[13] Heller JP, Kovarik F, Taber JJ. Improvement of CO_2 flood performance. Third Annual Report. Socorro (USA): New Mexico Institute of mining and Technology, New Mexico Petroleum Recovery Research Center; 1989

[14] Martin F, Heller J. Improvement of CO_2 flood performance—Quarterly report Jan.–March. PRRC. 1990;**1990**(2):90-20

[15] Zhang S, She Y, Gu Y. Evaluation of polymers as direct thickeners for CO_2 enhanced oil recovery. Journal of Chemical and Engineering Data. 2011;**56**(4):1069-1079

[16] Al Hinai NM et al. Experimental evaluations of polymeric solubility and thickeners for supercritical CO_2 at high temperatures for enhanced oil recovery. Energy and Fuels. 2018;**32**(2):1600-1611

[17] Al Hinai NM et al. Experimental study of miscible thickened natural gas injection for enhanced oil recovery. Energy and Fuels. 2017;**31**(5):4951-4965

[18] Bullen RS, Mzik J, Richard JP. Novel compositions suitable for treating deep wells. In: C.F. Ltd, editor. U.S Patent 4701270. 1987

[19] Sarbu T, Styranec TJ, Beckman EJ. Design and synthesis of low cost, sustainable CO_2-Philes. Industrial and Engineering Chemistry Research. 2000;**39**(12):4678-4683

[20] Cullick AS. Enhanced oil recovery using carbon dioxide. In: E.O. Corp, editor. U.S. Patent 4609043. 1986

[21] Liave FM, Chung FTH, Burchfield TE. Use of entrainers in improving mobility control of supercritical CO_2. SPE Reservoir Engineering. 1990;**5**(01):47-51

[22] Holtmyer MD, Hunt CV. Non-aqueous viscosified carbon dioxide and method of use. In: D. Halliburton Company, Okla, editor. U.S Patent 4964467. 1990

[23] Terry RE et al. Polymerization in supercritical CO_2 to improve CO_2/oil mobility ratios. In: SPE International Symposium on Oilfield Chemistry. San Antonio, Texas: Society of Petroleum Engineers; 1987

[24] Harris TV, Irani CA, Pretzer WR. Enhanced oil recovery using CO_2 flooding. In: C.R.a.T. Co, editor. U.S Patent 4913235A. 1990

[25] Bae JH. Viscosified CO_2 process: Chemical transport and other issues. In: SPE International Symposium on Oilfield Chemistry. San Antonio, Texas: Society of Petroleum Engineers; 1995

[26] Bae J, Irani C. A laboratory investigation of viscosified CO_2 process. SPE Advanced Technology Series. 1993;**1**(01):166-171

[27] Williams LL, Rubin JB, Edwards HW. Calculation of Hansen solubility parameter values for a range of pressure and temperature conditions, including the supercritical fluid region. Industrial and Engineering Chemistry Research. 2004;**43**(16):4967-4972

[28] Cai S. Study of CO_2 Mobility Control Using Cross-Linked Gel Conformance Control and CO_2 Viscosifiers in Heterogeneous Media. College Station, USA: Texas A & M University; 2011

[29] Al Yousef Z. Study of CO_2 Mobility Control in Heterogeneous Media Using CO_2 Thickening Agents. College Station, USA: Texas A & M University; 2012

[30] O'Brien MJ et al. Anthraquinone siloxanes as thickening agents for supercritical CO_2. Energy and Fuels. 2016;**30**(7):5990-5998

[31] Shen Z et al. CO_2-solubility of oligomers and polymers that contain the carbonyl group. Polymer. 2003;**44**(5):1491-1498

[32] McClain J, et al. Characterization of polymers and amphiphiles in supercritical CO_2 using small angle neutron scattering and viscometry. In: Proceedings of 1996 Spring Meeting of ACS, Division of Polymeric materials. AMER Chemical SOC, NW, WASHINGTON, DC 20036; 1996

[33] Wang Y et al. Design and evaluation of nonfluorous CO_2-soluble oligomers and polymers. The Journal of Physical Chemistry B. 2009;**113**(45):14971-14980

[34] Jianhang X. Carbon Dioxide Thickening Agents for Reduced CO_2 Mobility. Pittsburgh: University of Pittsburgh; 2003

[35] Xu J, Enick RM. Thickening carbon dioxide with the fluoroacrylate-styrene copolymer. In: SPE Annual Technical Conference and Exhibition. New Orleans, Louisiana: Society of Petroleum Engineers; 2001

[36] Tapriyal D. Design of Non-Fluorous Carbon Dioxide Soluble Compounds. Pittsburgh, USA: University of Pittsburgh; 2009

[37] Kilic S et al. Influence of tert-amine groups on the solubility of polymers in CO_2. Polymer. 2009;**50**(11):2436-2444

[38] Potluri VK et al. The high CO_2-solubility of per-acetylated α-, β-, and γ-cyclodextrin. Fluid Phase Equilibria. 2003;**211**(2):211-217

[39] Hong L, Thies MC, Enick RM. Global phase behavior for CO_2-philic solids: The CO_2+ β-d-maltose octaacetate system. The Journal of Supercritical Fluids. 2005;**34**(1):11-16

[40] Potluri VK et al. Peracetylated sugar derivatives show high solubility in liquid and supercritical carbon dioxide. Organic Letters. 2002;**4**(14):2333-2335

[41] Kilic S et al. Phase behavior of oxygen-containing polymers in CO_2. Macromolecules. 2007;**40**(4):1332-1341

[42] Drohmann C, Beckman EJ. Phase behavior of polymers containing ether groups in carbon dioxide. The Journal of Supercritical Fluids. 2002;**22**(2):103-110

[43] Tapriyal D et al. Poly(vinyl acetate), poly (1-O-(vinyloxy) ethyl-2,3,4,6-tetra-O-acetyl-β-d-glucopyranoside) and amorphous poly(lactic acid) are the most CO_2-soluble oxygenated hydrocarbon-based polymers. The Journal of Supercritical Fluids. 2008;**46**(3):252-257

[44] Heller JP, Kovarik FS. Improvement of CO_2 flood performance. Fourth Annual Report for the Period of Oct. 1, 1987–Sept. 30, 1988. New Mexico Petroleum Recovery Research Centre; 1988

[45] Babu SS, Praveen VK, Ajayaghosh A. Functional π-gelators and their applications. Chemical Reviews. 2014;**114**(4):1973-2129

[46] Terech P, Weiss RG. Low molecular mass gelators of organic liquids and the properties of their gels. Chemical Reviews. 1997;**97**(8):3133-3160

[47] Escuder B, Miravet JF. Silk-inspired low-molecular-weight organogelator. Langmuir. 2006;**22**(18):7793-7797

[48] Terech P et al. structural variations in a family of orthodialkoxyarenes organogelators. Journal of Colloid and Interface Science. 2006;**302**(2):633-642

[49] Trickett K et al. Rod-like micelles thicken CO_2. Langmuir. 2009;**26**(1):83-88

[50] John C. Structure of cobalt aerosol-OT reversed micelles studied by small-angle scattering methods. Journal of the Chemical Society, Faraday Transactions. 1994;**90**(17):2497-2504

[51] Hirst AR et al. Low-molecular-weight gelators: Elucidating the principles of gelation based on Gelator solubility and a cooperative self-assembly model. Journal of the American Chemical Society. 2008;**130**(28):9113-9121

[52] Hanabusa K et al. Two-component, small molecule gelling agents. Journal of the Chemical Society, Chemical Communications. 1993;**18**:1382-1384

[53] Stals PJ et al. The influence of oligo (ethylene glycol) side chains on the self-assembly of benzene-1, 3, 5-tricarboxamides in the solid state and in solution. Journal of Materials Chemistry. 2009;**19**(1):124-130

[54] Dandge DK et al. Associative organotin polymers. 2. Solution properties of symmetric trialkyltin fluorides. Journal of Macromolecular Science: Part A—Chemistry. 1989;**26**(10):1451-1464

[55] Shi C et al. Semi-fluorinated trialkyltin fluorides and fluorinated telechelic ionomers as viscosity-enhancing agents for carbon dioxide. Industrial and Engineering Chemistry Research. 2001;**40**(3):908-913

[56] Iezzi A, Enick R, Brady J. Direct viscosity enhancement of carbon dioxide. In: Supercritical Fluid Science and Technology. American Chemical Society; 1989. pp. 122-139

[57] Johnston KP, Penninger JML. Supercritical fluid science and technology. In: ACS Symposium Series. Vol. 406. American Chemical Society; 1989. 564p

[58] Fieser LF et al. Napalm. Industrial and Engineering Chemistry. 1946;**38**(8):768-773

[59] Hughes M. Napalm: An American biography by Robert M. Neer. War in History. 2014;21(4):560-562

[60] Mysels KJ. Napalm. Mixture of aluminum disoaps. Industrial and Engineering Chemistry. 1949;**41**(7):1435-1438

[61] Koide K. Latex composition comprising a cross-linking agent and molded product thereof. U.S. Patent 20080227913A1. 2008

[62] Enick R. The effect of hydroxy aluminum disoaps on the viscosity of light alkanes and carbon dioxide. In: SPE International Symposium on Oilfield Chemistry. Anaheim: California Society of Petroleum Engineers; 1991

[63] Lewis P. An Attempt to Increase the Viscosity of CO_2 with Metallic Stearates. Pittsburgh, USA: University of Pittsburgh; 1990

[64] Iezzi A et al. 'Gel' formation in carbon dioxide-semifluorinated alkane mixtures and phase equilibria of a carbon dioxide-perfluorinated alkane mixture. Fluid Phase Equilibria. 1989;**52**:307-317

[65] Gullapalli P, Tsau JS, Heller JP. Gelling behavior of 12-hydroxystearic acid in organic fluids and dense CO_2. In: SPE International Symposium on Oilfield Chemistry. San Antonio, Texas: Society of Petroleum Engineers; 1995. p. 13

[66] Shi C et al. The gelation of CO_2: A sustainable route to the creation of microcellular materials. Science. 1999;**286**(5444):1540-1543

[67] Paik IH et al. Fiber formation by highly CO_2-soluble bisureas containing peracetylated carbohydrate groups. Angewandte Chemie International Edition. 2007;**46**(18):3284-3287

[68] Eastoe J et al. Variation of surfactant counterion and its effect on the structure and properties of aerosol-OT-based water-in-oil microemulsions. Journal of the Chemical Society, Faraday Transactions. 1992;**88**(3):461-471

[69] Eastoe J et al. Structures of metal bis (2-ethylhexylsulfosuccinate) aggregates in cyclohexane. The Journal of Physical Chemistry. 1993;**97**(7):1459-1463

[70] Cummings S et al. Design principles for supercritical CO_2 viscosifiers. Soft Matter. 2012;**8**(26):7044-7055

[71] Eastoe J et al. Water-in-CO_2 microemulsions studied by small-angle neutron scattering. Langmuir. 1997;**13**(26):6980-6984

[72] Eastoe J, Gold S, Steytler DC. Surfactants for CO_2. Langmuir. 2006;**22**(24):9832-9842

[73] Eastoe J et al. Effects of fluorocarbon surfactant chain structure on stability of water-in-carbon dioxide microemulsions. Links between aqueous surface tension and microemulsion stability. Langmuir. 2002;**18**(8):3014-3017

[74] Liu ZT, Erkey C. Water in carbon dioxide microemulsions with fluorinated analogues of AOT. Langmuir. 2001;**17**(2):274-277

[75] Doherty MD et al. Small molecule cyclic amide and urea based thickeners for organic and sc-CO_2/organic solutions. Energy and Fuels. 2016;**30**(7):5601-5610

[76] Koottungal L. Survey: Miscible CO_2 continues to eclipse steam in US EOR production. Oil and Gas Journal. 2014;**112**(4):78-91

[77] Jaber AK, Awang MB. Field-scale investigation of different miscible CO_2-injection modes to improve oil recovery in a clastic highly heterogeneous reservoir. Journal of Petroleum Exploration and Production Technology. 2017;7(1):125-146

[78] Kirby CF, McHugh MA. Phase behavior of polymers in supercritical fluid solvents. Chemical Reviews. 1999;**99**(2):565-602

[79] C AR Jr. Oil Recovery process by miscible displacement. U.S. Patent 3444930. 1969

[80] Dauben DL, Reed J, Shelton J, Lyman Y. Recovery of oil with viscous propane. U.S. Patent 3570601. 1971

[81] Henderson J, Meyer W, Taber J. Miscible drive secondary oil recovery process. U.S. Patent 3330345. 1967

[82] Dhuwe A. Thickeners for Natural Gas Liquids to Improve the Performance in Enhanced Oil Recovery and Dry Hydraulic Fracking. University of Pittsburgh; 2016

[83] Dhuwe A et al. Assessment of solubility and viscosity of ultra-high molecular weight polymeric thickeners in ethane, propane and butane for miscible EOR. Journal of Petroleum Science and Engineering. 2016;**145**:266-278

[84] Dindar C. High-pressure viscosity and density of polymer solutions at the critical polymer concentration in near-critical and supercritical fluids. In: Chemical Engineering. Pittsburgh, USA: Virginia Polytechnic Institute and State University; 2001

[85] Dindar C, Kiran E. High-pressure viscosity and density of polymer solutions at the critical polymer concentration in near-critical and supercritical fluids. Industrial and Engineering Chemistry Research. 2002;**41**(25):6354-6362

[86] Wu MM. Process for manufacturing olefinic oligomers having lubricating properties. U.S. Patents 4827073A. 1989

[87] Zolper T et al. Lubrication properties of polyalphaolefin and polysiloxane lubricants: Molecular structure–tribology relationships. Tribology Letters. 2012;**48**(3):355-365

[88] Dhuwe A et al. Small associative molecule thickeners for ethane, propane and butane. The Journal of Supercritical Fluids. 2016;**114**:9-17

[89] Lee J et al. Polymeric and small molecule thickeners for CO_2, ethane, propane and butane for improved mobility control. In: SPE Improved Oil Recovery Conference. Tulsa, Oklahoma, USA: Society of Petroleum Engineers; 2016

[90] Smith KW, Persinski LJ. Hydrocarbon gels useful in formation fracturing. In: C. Inc, editor. U.S Patents 5571315A. 1996

[91] Van Den Berghe EV, Van Der Kelen GP. A study of the chemical bond in (CH3)nSn(SCH3)4−n (n=0, 1, 2, 3) by nmr (1H, 119Sn, 13C) spectroscopy. Journal of Organometallic Chemistry. 1971;**26**(2):207-213

[92] Smith KW, Persinski LJ. Hydrocarbon gels useful in formation fracturing. In: C. Inc, editor. U.S Patent 5614010. 1997

[93] Taylor RS, Funkhouser GP. Methods and compositions for treating subterranean formations with gelled hydrocarbon fluids. In: H.E.S. Inc, editor. U.S. Patent 6511944. 2003

[94] Taylor RS, Funkhouser GP. Methods and compositions for treating subterranean formations with gelled hydrocarbon fluids. In: H.E.S. Inc, editor. U.S. Patent 7314850. 2008

[95] Taylor RS, et al. Compositions and methods for treating subterranean formations with liquefied petroleum gas. In: C.U.I.E.S. Inc, editor. U.S. Patent 7341103. 2008

[96] George M, Funkhouser GP, Weiss RG. Organogels with complexes of ions and phosphorus-containing amphiphiles as gelators. Spontaneous gelation by in situ complexation. Langmuir. 2008;**24**(7):3537-3544

[97] Delgado E, Keown B. Low volatile phosphorous gelling agent. U.S. Patents 8377854B2. 2013

[98] Taylor R et al. Optimized gas-well stimulating using CO_2-miscible, viscosified hydrocarbon fracturing fluids. In: SPE Gas Technology Symposium. Calgary, Alberta, Canada: Society of Petroleum Engineers; 2002. p. 8

[99] Taylor RS et al. Rheological evaluations of CO miscible hydrocarbon fracturing fluids. In: Canadian International Petroleum Conference. Calgary, Alberta: Petroleum Society of Canada; 2005. p. 8

Chapter 5

Potential of Low-Salinity Waterflooding Technology to Improve Oil Recovery

Hisham Ben Mahmud, Shattia Arumugam and Walid Mahmud

Abstract

Low-salinity waterflooding (LSWF) is a potential new method for enhanced oil recovery (EOR) in sandstone and carbonate rock formations. LSWF approach has gained an attention in the oil and gas industry due to its potential advantages over the conventional waterflooding and other chemical EOR technologies. The efficiency of waterflooding process is effected via reservoir and fluid parameters such as formation rock type, porosity, permeability, reservoir fluid saturation and distribution and optimum time of water injection. Combined effect of these factors can define the ultimate recovery of hydrocarbon. The main objective of this chapter is to review the mechanism of LSWF technique in improving oil recovery and the mechanism under which it operates. Various laboratory studies and few field applications of LSWF in recent years have been presented mainly at the lab scale. Also it will explore numerical modeling developments of this EOR approach.

Keywords: low salinity, enhanced oil recovery, reservoir and fluid properties, LSWF modeling development

1. Primary recovery

The hydrocarbon fluid, crude oil, is a naturally occurring non-renewable resource, and it is one of the fossil fuels which the world's economy mostly depends on. Crude oil is composed of hydrocarbon deposits and other organic materials that can be refined and processed further to produce various chemical products. The production process of hydrocarbon fluid is divided into three stages, which are, namely, primary, secondary and tertiary recovery. The primary recovery of the crude oil is driven via the natural source of energy available in the reservoir such as solution gas drive, aquifer drive, gravity drainage, gas cap drive and rock and fluid expansion. Moreover, the extracted oil by artificial lift technologies (i.e. gas lifts, electrical submersible pump (ESP)) is considered a primary recovery. This stage of recovery is very limited to only 5–15% of the original oil in place (OOIP) which is produced [1]. When the available natural energy in the reservoir decreases overtime resulting in a significant drop of oil production, thus an external

IntechOpen

energy must be added to the reservoir to maintain the reservoir pressure in order to produce additional oil. This stage is called secondary oil recovery, which utilizes various mechanisms including gas injection and waterflooding into the reservoir to force and displace the remaining residual oil. This process is typically successful in producing around 30% of the oil reserves after natural depletion, leaving 50–80% of oil still unrecovered [2].

The last stage of hydrocarbon recovery is known as enhanced oil recovery (EOR), which uses different practices such as chemical flooding, miscible flooding and thermal methods to extract the hydrocarbon fluid left behind the primary and secondary recovery. EOR has the capability of increasing oil recovery up to 75% of OOIP by improving the mobility of oil via modifying fluid properties [3]. Some examples of EOR techniques implemented in the oil industry are polymer flooding, steam injection, alkaline flooding, in situ combustion and modified waterflooding.

2. Secondary recovery using waterflooding

Conventional waterflooding is a secondary oil recovery approach that consists of water injection to improve the oil production from the subsurface. It is typically performed after the primary recovery which utilizes the natural energy available in the reservoir. The main purpose of secondary recovery is to displace hydrocarbons towards the production wells while maintaining the reservoir pressure.

The improved oil production using waterflooding was first discovered in early 1865 following an accidental flooding of water in Pithole City, Pennsylvania. This was the result of leaks from surface water and shallow water which entered the drilled holes. It is revealed that the oil recovery factor by waterflooding is significantly high compared to the natural depletion. The first applied application of waterflooding was attempted in Pennsylvania's Bradford field, in 1924, which then grew and was widely applied in many fields in the subsequent decades [4]. Following nationwide waterflooding implementation in petroleum industry, many attempts were conducted in understanding mechanism, planning and optimizing the process. Due to its simplicity and reliability, the waterflooding technique has been worldwide implemented and been considered for most of conventional oil reservoirs to extract more hydrocarbon after the primary recovery process.

The conventional waterflooding process involves water injection into the reservoir formation in which the process is generally done with consideration of the economic factors and also based on the water compatibility with the present reservoir brine to avoid formation damage. However, in the early 1990s, a number of researchers experimentally investigated the effect of water composition and found that it plays a significant role in the oil recovery. After this, the potential of low-salinity waterflooding (LSWF) in EOR applications was observed and developed by Morrow and his coinvestigators [5].

Furthermore, extensive water coreflood experiments have been conducted and addressed the benefits of low salinity in the EOR process. Most of these experiment results showed that when the injected water salinity is lower than the formation water salinity, a higher oil recovery up to 40% is achieved for both secondary and tertiary recoveries [6]. However, LSWF has gained vast interest in the petroleum industry due to its practical advantages compared to other chemical EOR methods. LSWF is an emerging EOR technology, and it has a promising future since half of the world's petroleum originates from sandstone reservoirs.

Next section will cover LSWF process in details.

3. Enhanced oil recovery using low-salinity waterflooding

3.1 Concept of mobility ratio

As seen earlier waterflooding involves the injection of water into the reservoir is by far the most widely utilized solution for improving oil recovery. Practically any oil field which does not have an aquifer support will be considered for waterflooding to maintain the reservoir pressure and improve production rate [7]. A reservoir's compliance or suitability for waterflooding to increase oil production can be done by evaluating formation and fluid parameters such as formation rock type, porosity, permeability, saturation and distribution of reservoir fluids and optimum time of water injection [8]. Combined effect of these factors can determine the ultimate recovery of hydrocarbon and its economic revenues depicting the viability of carrying out waterflooding for a specific reservoir condition.

The efficiency of a flooding process can be qualitatively evaluated via defining the mobility ratio, especially end-point mobility ratio. As the term implies, end-point mobility ratio is measured at the end-point saturation of a single-phase fluid and can be written as shown in Eq. (1).

$$M = \frac{\lambda_{water}}{\lambda_{oil}} = \frac{k_{rw}/\mu_w}{k_{ro}/\mu_o} = \frac{k_{rw}\mu_o}{k_{ro}\mu_w} \tag{1}$$

where,

M end-point mobility ratio
λ_{water} water mobility (mD)
λ_{oil} oil mobility (mD/cP)
k_{rw} relative permeability of water (mD)
k_{ro} relative permeability to oil (mD)
μ_o oil viscosity (cP)
μ_w water viscosity (cP)

When the value of end-point mobility ratio is less than one, it indicates that the performed flooding is stable, while for a value of more than one, flooding process is unstable due to a phenomenon known as 'viscous fingering'. Considering an oil–water system, stable flooding signifies that oil displacement will effectively take place if the injected water behaves like a piston and pushes oil to the intended point [9]. On the other hand, viscous fingering refers to early and continuous breakthrough of injected fluid as a result of large difference in viscosity between water and oil phase. A higher value of mobility ratio implies reducing waterflooding effectiveness as the volumetric sweep efficiency reduces [10].

3.2 Concept of wettability

In understanding the process of oil recovery, formation wettability knowledge is important as it describes the reservoir performance via defining the fluid flow and distribution. Being one of the most significant factors of LSWF, it is important to comprehend wettability phenomenon appropriately to prevent any incorrect assumptions which may lead to permanent formation damage.

Wettability can be defined in a system which consists of two immiscible fluids in contact with a solid surface (rock). In the presence of such system, wettability can be described as the tendency of one fluid to adhere to the rock surface to be in contact with one fluid than the other. In the situation of two-phase immiscible fluid, one fluid attaches strongly to the rock surface while displacing the other fluid [11]. In reference to wettability concept, fluids can be classified into wetting or non-wetting fluid. A wetting fluid balances its forces and adheres to the rock at a specific contact angle, while the non-wetting fluid will have minimal or no contact with the rock.

In a reservoir system, when the reservoir fluids present in the porous medium are oil and water, wettability can be described as the preference of the rock to be in contact with either oil or water or both fluids. In an oil field, reservoir rock wettability is typically described as either water-wet or oil-wet. If the reservoir is described to be water-wet, it means that the water phase is retained on the pore wall either small or large by capillary pressure, while the oil phase occupies the pore space [12]. On the other hand, an oil-wet reservoir implies that oil phase is adhered to the rock pores, while water phase occupies the centre of pores. There are a number of techniques in which wettability can be quantitatively measured by the determination of contact angle, the Amott method and US Bureau of Mines (USBM) method [13]. The surface energies of a water, oil and solid system can be expressed by Young's equation; Eq. (2) can be deduced.

$$\sigma_{ow} \cos\theta = \sigma_{os} - \sigma_{ws} \tag{2}$$

where,

σ_{ow} oil and water interfacial energy (dyne/cm)
σ_{os} oil and solid interfacial energy (dyne/cm)
σ_{ws} water and solid interfacial energy (dyne/cm)
θ contact angle (degree)

As a typical practice, the contact angle measurement is performed through the aqueous phase, and it identifies wettability as seen in **Figure 1**. For a reservoir rock containing only oil and water, a contact angle of less than 90° indicates that the reservoir rock is water-wet, but when the contact angle is more than 90°, it denotes the reservoir rock is oil-wet. Moreover, a strong water-wet rock system can present when the fluid-rock contact angle approaches to 0°, while a strong oil-wet rock can be described when the contact angle approaches to 180° [13]. When both fluids (oil, water) are in contact with the rock surface, the reservoir rock can be described to be in intermediate/neutral-wet condition.

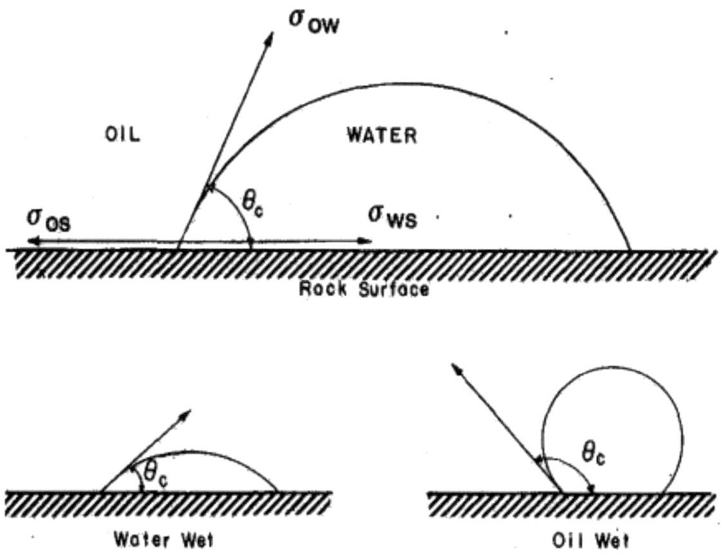

Figure 1.
Wettability of the crude oil/brine/rock system [13].

Wettability could greatly affect the reservoir rock petrophysical properties, for instance, residual saturation, capillary pressure, relative permeability and end point of relative permeability curves [14]. The crossover point and the relative permeability end points of wetting and non-wetting phases are related to wettability as shown in **Figure 2**. For a strong water-wet rock, the relative permeability curves will crossover at a wetting phase saturation point of greater than 0.5, while a strong oil-wet rock relative permeability curves will crossover at a wetting phase saturation point of lesser than 0.5. A crossover point at a saturation of 0.5 and equal end points of relative permeability curve imply that the reservoir rock is in neutral-wet condition [16].

The alteration in reservoir rock wettability may possibly occur naturally during production, or it can be modified using thermal or chemical method. Initially, most of the reservoir formations are in the state of strong water-wet due to the deposition process that saturates reservoir completely with water [17]. The migration of hydrocarbon fluid (especially oil) may cause a change in the rock wettability to oil-wet, or the rock may maintain its wettability as water-wet. Mugele et al. [18] reported that the rock wettability can be altered to be more water-wet via the adsorption of divalent cations, subsequently allowing better mobilization of oil for production. In addition, several studies strongly suggest that the adsorption of divalent cations such as Ca^{2+} and Mg^{2+} changes the initial rock wettability towards more water-wet [19]. This is an important observation that describes the LSWF mechanism which results in incremental oil recovery and will be deeply discussed in the upcoming sections.

3.3 Role of crude oil, brine and reservoir rock on wettability

It is commonly known that oil reservoirs can have various alterations of wettability depending on the oil and rock and also the composition and amount of the brine phase. Crude oil/brine/rock (COBR) interactions are utilized to produce wetting conditions in laboratory core samples which are more representative of wetting in an oil reservoir than either a strongly oil-wet or water-wet. Anderson [20] stated that the original strong water-wetness of most reservoir minerals can be changed via the polar compounds adsorption and/or the deposition of organic matter that was initially in the crude oil. He also reported the surface-active agents present in

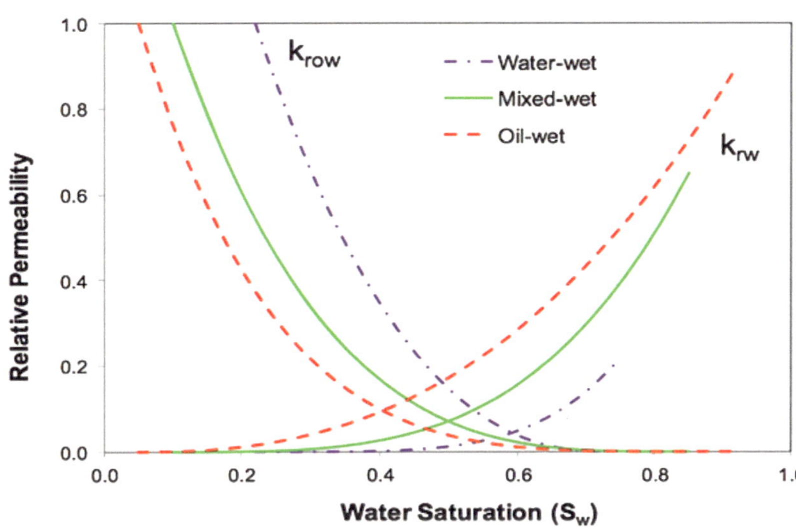

Figure 2.
Effect of wettability on relative permeability curve [15].

the oil that are widely believed to be polar compounds that comprise nitrogen, oxygen and/or sulfur. Such compounds have a polar and a hydrocarbon end. He noted that the polar end is adsorbed on the rock surface, exposing the hydrocarbon end and making the rock surface more oil-wet. Furthermore, to the oil composition, the degree of wetting conditions which is altered by these surfactants can also be determined by various factors that are the temperature, pressure, mineral surface and brine chemistry, including pH and ionic composition.

Buckley et al. [21] found that some factors are affecting the rock wetting upon contact with crude oil, including composition of both the oil and brine, temperature and duration of aging in oil and initial water saturation. The chemistry of the brine phase was noticed to change the rocks wettability, which shown that the brine pH is also significant in verification of wettability and other interfacial properties of the COBR system. They also indicated that similar factors affect measurements of contact angle using asphaltic crude oil and two sodium chloride solutions of different pH to alter the wetting of two core formations. They noted in cores that there are further complexities in wetting related to rough surfaces, converging and diverging pore shapes and heterogeneous mineralogy. Superimposed on all these sources of heterogeneity in the porous media is the ability of crude oil components to adsorb onto mineral surfaces and change their wetting properties. They summarized that changing pH of aqueous phase affects initial water saturation (Swi) in cores with low-ionic-strength NaCl solutions. Moreover, pore coatings may control wetting alteration in the porous media thus the COBR interactions are not similar to predicted values.

Tang and Morrow [22] performed a study on the effect of salinity and oil composition on wettability behaviour and observed that there are some possible mechanisms by which COBR interactions control wettability and efficiency of oil recovery. They noted that acid–base interactions can define oil-brine and brine-solid surface charges, and also direct adsorption from crude oil onto a dry surface is ascribed to polar interactions. It is noticed that alterations in wettability resulting from instability of a water film take place almost instantaneously. Furthermore, adsorption by ion binding takes place through attachment of polar components in the crude oil to specific surface sites on the solid surface by multivalent ions. Surface precipitation is recognized by conditions of poor solvency of asphaltenes in the oil phase. They noted that molecular association, including ion binding of crude oil components at the interface of oil–water, could promote lateral aggregation and the formation of an organic mat that may still be largely separated from the solid surface by a thin water film. They concluded that wetting variations of a rock induced by crude oil are related to changes in solvency of the crude oil with respect to its heavy polar components.

Cuiec [23] explained many experimental condition contributions to understand oil/rock interactions, which are responsible on the occurrence of oil-wet reservoirs. He noticed that some intermediate fractions of crude oil samples might change the rocks' surface properties depending on the type of formation. He also reported the role of asphaltenes in crude oil/solid interactions through correct correlation between wettability and asphaltene content for a set of reservoirs.

Clementz [24] observed that clay mineral properties are irreversibly altered by adsorption of heavy crude ends. Therefore, this alteration causes reduction in rock sensitivity to injected fluids since rock wettability is altered from water-wet to neutral. He noted further that adsorption of surfactants is decreased, while rock property measurements are changed from irreversibly altered cores. The extent of this interaction is subjected to the type of clay minerals in the rock, the composition of heavy crude ends and the interaction environment. He had seen that after the adsorption took place, a clay-organic complex is formed which is hydrophobic and very stable. The development of swelling clays is decreased, and clay surface area and cation exchange capacity are decreased as well. As a result, he revealed that the practical implication of these problems indicates some sort of spotted coverage of the surface and leads to discuss the role of an adsorbed layer on the rock wettability. The existence of this surface film would certainly affect oil recovery in a rock system. He recommended that it cannot be stated a priori that oil recovery from the rock having a neutral wettability will be less than that from the water-wet system. In addition, he reported that the recovery would most undoubtedly rely on the nature of drive fluid and its interaction with reservoir minerals. Finally, he concluded that it is very important to identify potential alterations to core properties, which take place as a result of core handling. The absorbed layer can either be formed or damaged, and any rock property measurements, which count on the nature of the clay minerals present, would be affected by the existence of the adsorbed layer.

Brown and Neustadter [25] studied the wettability performance of oil/water/silica systems based on contact angle measurements. They reported that the recovery efficiency of crude oil from a porous media using water displacement depends on the rock wettability. This conclusion was based on the relative tendency of both the aqueous and oil phase to coat the solid surface and, therefore, to occupy the rock pores under the action of capillary forces. They also found that the existence of monovalent ions can affect wetting conditions by suppressing the charge effects with no effect at low concentration of NaCl and behave differently at high concentration and extremes of pH. By contrast to the effect of monovalent ions, they have seen that the seawater can provide a strongly oil-wetted surface, independent of pH, even when it is diluted with distilled water. Furthermore, they noticed the same performance with synthetic seawater that was applied to determine if the original seawater can affect the wettability by reason of its protein or other nonsalt components. The results obtained with NaCl were as expected considering the charge interaction model, but using seawater indicated that divalent ions had a very noticeable effect. In addition, they conducted a study to examine the effect of divalent ions and observed that there are specific interactions between different oil crude samples and the divalent ions can take place. They suggested that alterations in wettability are not contributed to general electrical double-layer compression at low electrolyte concentrations. They also observed that divalent ions can create bridges between specific petroleum crude surfactant species and the silica surface. In conclusion, for systems of crude oil/silica/distilled water, the wettability strongly depends on pH due to activation of the crude oil surfactants by the aqueous phase. When pH value is high, the water film can stabilize between the oil film and the silica surface by charge repulsion, while at low pH values, charge attraction between the positively charged crude oil surfactants and the negative rock surface promotes wetting oil surface. Moreover, crude oil/rock systems, which are preferentially oil-wet, show large hysteresis between the advancing and receding angles. Due to the low solubility of the crude oil surfactants in the aqueous phase, the crude oil has to adhere to the surface before the surfactants, which promote oil wettability, can be adsorbed.

3.4 Low-salinity enhanced oil recovery

The recovery of hydrocarbon fluid from subsurface formations is a complex process that associates multiple length scales. In typical water flooding process, seawater is injected on a macroscopic scale into the reservoir formation which carries hydrocarbon fluid (oil) to sweep the oil away from injection wells towards production wells. On the microscopic scale, the injected water displaces the oil in the porous reservoir formation in a typical microfluidic two-phase flow of water and oil. Due to the large ratio of surface-to-volume, the efficiency of this process of microfluidic is strongly influenced by the surrounding porous rock matrix wettability. The rock formation ingredient, typically sandstone or limestone, is made of naturally hydrophilic material. Throughout millions of years of exposure to petroleum fluid, a layer of organic material adsorbed onto the surfaces and thus rendered the rock more hydrophobic. These molecular scale adsorption developments (as seen in **Figure 3**) thereby impede the oil displacement via water phase in the hydrophobized rock pores. This is understood to be one of the major causes why the secondary oil recovery is a rather inefficient process that leaves more than 50% of the oil unrecovered in the reservoir [26].

Since the 1970s, oil operator companies have examined several techniques to improve the low recovery rate by injecting various additives combined with the water aqueous phase. More recently, it was found that the recovery rate can be improved by desalinating the seawater before injecting it into the reservoir. The following sections will explore experiment and numerical developments of low water salinity approach.

3.5 Experimental observation of low salinity

Upon the observation that the injection of freshwater in sandstone reservoirs reduces the oil recovery due to clay swelling, extensive laboratory researches were conducted in the 1940s to evaluate the influence of fluid's physical and chemical properties on oil recovery [27]. Reiter [28] observed that low-salinity waterflooding obtained an additional oil recovery of 21.3% more than higher-saline floods conducted on Nacatoch sandstone cores. Further investigation by [29] on the effect of salt water on oil recovery containing clays concluded that the pressure drop across cores and oil recovery increased as the concentration of salt-in injection water decreased.

The true EOR potential of LSWF was recognized by Morrow and his experimental co-workers from the studies conducted on the effect of wettability on oil recovery via waterflooding [30]. They confirmed that the composition of injection

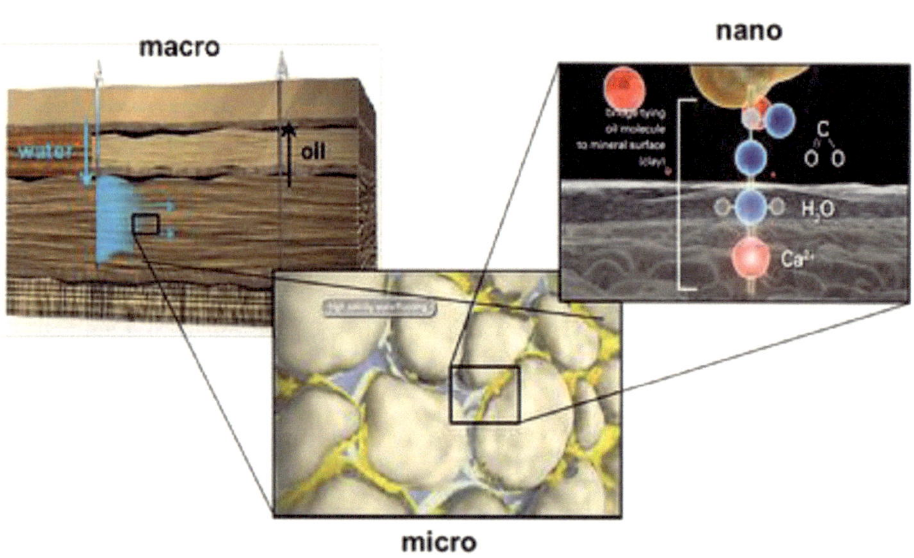

Figure 3.
Molecular scale adsorption processes [26].

brine affects oil recovery, but the amount of recovery depends on the conditions of crude oil, injection brine and reservoir rock. Further investigation by [30] on LSWF observed that the oil recovery obtained from conventional water flooding with high-salinity brine was similar to that of LSWF when initial formation water saturation was zero. It was concluded that the positive effects of low salinity could only be obtained with the existence of connate water, and the salinity of connate water plays a significant role in the amount of oil recovery. It was observed from the experiment that additional oil recovery could be achieved when the injection water salinity is relatively lower than formation water salinity.

However, this was not observed in all the studies conducted on the effect of brine composition on oil recovery. Zhang [31] presented their report showing that the injection of low-salinity brine on Berea sandstone resulted in lower recovery than that of high-salinity waterflooding, although more cases showed better recovery with the application of low-saline brine injection. The salinity level of 1500 ppm NaCl showed higher oil recovery, while the injection of 8000 ppm had zero effect although both the salinity levels were below the connate water salinity. Nevertheless, in most of the published cases, LSWF showed positive benefits on oil recovery in sandstones.

Based on practices in the laboratory, Jerauld et al. [32] proposed that brine composed of 10–25% of connate water or salinity of 1000 to 2000 ppm will be an appropriate estimation in determining the composition of injection brine. A total of 214 and 188 laboratory scale studies conducted in the secondary mode and tertiary mode, respectively, which confirmed the positive effect as an increment of 5–20% in oil recovery was observed [33]. The application of LSWF evaluated in West Salym field, Russia, through coreflood tests on sandstone cores resulted in elevated oil recovery of 4% OOIP and 1.7% OOIP in the tertiary mode [34]. A summary of experimental research that has been conducted to study the benefits of LSWF in improving oil recovery is shown in **Table 1**.

Reference	Rock type	Injected water salinity (ppm)	Formation water salinity (ppm)	Mode of injection	Incremental oil recovery achieved (%)
[35]	Sandstone	1480 SW 1500 NaCl	29,690 FW	Secondary	29
				Tertiary	7–14
[36]	Sandstone	50–5500 SW	22,000 SW	Secondary	14–28
[37]	Berea sandstone	10,000 NaCl	40,000 NaCl	Secondary	5–6
				Tertiary	25–35
[38]	Berea and field sandstone	870–1140 SW	30,510 SW	Secondary	15
				Tertiary	0
[39]	Berea and field sandstone	1% FW	FW	Secondary	10–22
				Tertiary	2–6
[40]	Sandstone	1054.96–105.49 SW	105,496 FW	Secondary	9
				Tertiary	3
[34]	Sandstone	—	FW	Secondary	6
				Tertiary	5

Note: SC, standard conditions of temperature (60°F) and pressure (1 atm); RC, reservoir condition of temperature and pressure; SW, seawater; FW, formation water.

Table 1.
Summary of laboratory experimental observation on low-salinity waterflooding.

3.6 Proposed mechanism of low-salinity waterflooding

3.6.1 Fine migration

Clay is subjected to swelling when it is exposed to freshwater. Early studies conducted in understanding the mechanisms of low-salinity water injection are associated with clay swellings and fine migration. According to Sheng [41], fine mobilization takes place when the strength of ions in injected brine is lower than the concentration of critical aggregate lump. Divalent cations play a major role in determining the critical flocculation concentration. These cations stabilize clay by reducing the zeta potential and repulsive force. The injection of solution with a low-saline concentration destabilizes and disperses clay from the wall of pores causing it to flow with water. Clay that flows in water tends to stick on the surface of small pores resulting in the reduction of permeability. This phenomenon increases the sweep efficiency as water is forced to flow on new paths.

Jackson et al. [42] reported similar mechanism stating that low-salinity effect arises from the removal of mixed-wet fines from the surface of formation and accumulation of oil-wet fines at the oil–water interface. This increases oil recovery as it stimulates the mobilization of oil and changes the wettability of the formation towards more water-wet by exposing the water-wet surfaces beneath the stripped fines as denoted by **Figures 4–6**.

Based on their experimental studies, Tang and Morrow [43] reported that incremental oil recovery obtained through LSWF was due to fine mobilization especially kaolinite. The conclusion was made upon their observation that unfired Berea sandstone showed improvement on oil recovery during low-saline brine injection, while acidized Berea core did not show any improvements. It was also observed in their research that the increase in oil recovery was more significant for sandstones containing clays than clean core samples.

However, numerous LSWF carried out by British Petroleum (BP) on cores under reduced and reservoir condition in sandstones did not indicate any fine mobilization despite the increment in oil recovery [32]. Zeinijah et al. [44] reported that minimal to zero amount of clay production was observed during their experiment of flooding cores with low-saline brine. The variation in the composition of injected brine, minerals and lithology might be the reason for the conflicting findings.

3.6.2 Increase in pH and reduction in interfacial tension

Mcguire et al. [45] suggested that similar to the mechanism of alkaline flooding, increase in the pH value and reduction in interfacial tension (IFT) could be one of the LSWF mechanisms. The exchange of adsorbed sodium ions with hydrogen ions in water results in elevation of pH. Various experiments conducted by researches indicated an increase in pH value of about pH 2 to pH 4 upon the injection of low-saline brine. In situ surfactant that lowers oil or water IFT is produced when organic acids in the crude oil react under high-pH conditions [46]. The formation of surfactants and reduction of IFT forms either oil–water emulsion or water–oil emulsions which results in the improvement of water sweep efficiency [33].

However, in several cases, in the injection of low-saline water, the pH value was lower than 7, and in some cases pH remains unchanged. Zhang et al. [31] reported that no obvious relationship was observed between effluent pH and oil recovery, and only slight change in pH was observed during low-salinity injection. The measured IFT value was above 10 mN/m with pH less than 9 during the course of LSWF which was concluded to be very low to reduce residual oil saturation.

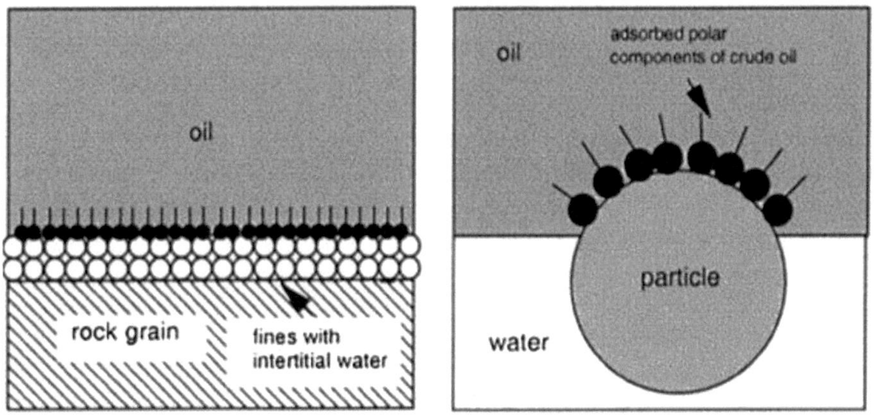

Figure 4.
Formation of mixed-wet fines due to adsorption of polar components from crude oil [42].

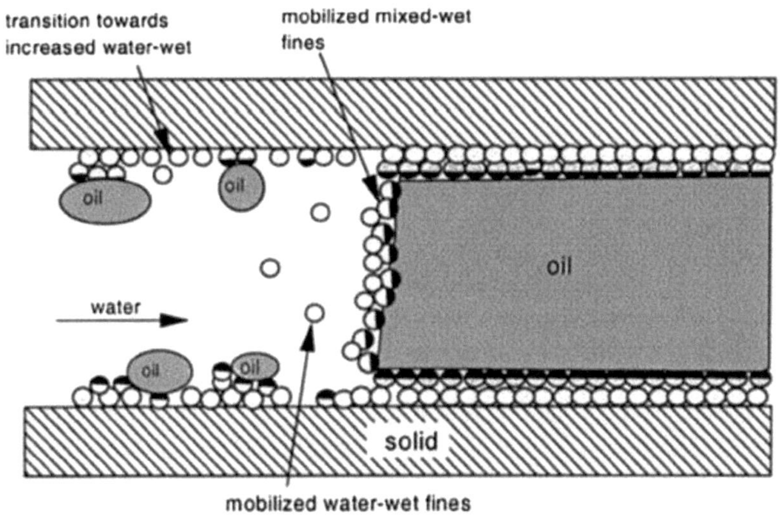

Figure 5.
Stripping of mixed-wet fines from pores during waterflooding [42].

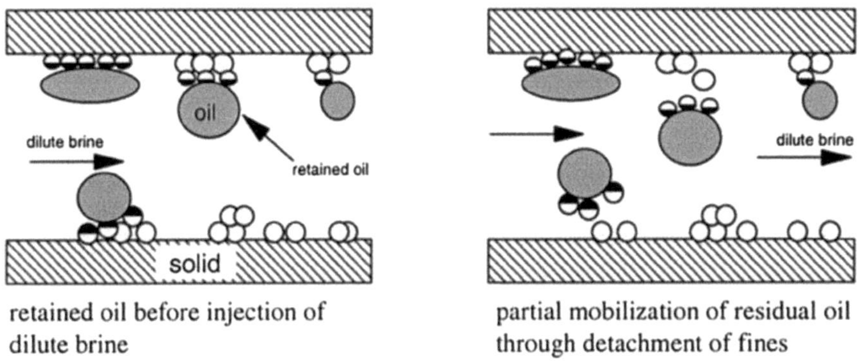

Figure 6.
Mobilization of trapped oil [42].

A chemical mechanism proposed by Austad [46] signifies the role of clay at low pH values. Thermodynamic chemical equilibrium that initially exists at reservoir condition at low pH increases the adsorption of anions and cations onto the clay surface. The injection of low-salinity brine disturbs this chemical equilibrium causing reaction between rock and brine to occur especially during the presence of Ca^{2+} ions. Compensating for the loss of cation into the low-saline water, H^{+} ions reacts and increases the pH near to the clay surface. Thus, an increase in pH is introduced by the tendency of low-salinity brine in changing the chemical structures initially present.

3.6.3 Multicomponent ion exchange

Multicomponent ion exchange (MIE) mechanism describes the alteration in the wettability of reservoir rock towards more water-wet due to the release of oil particles from the clay surfaces. Low-salinity water expands the double layer and eases the process of desorption of divalent ion on oil bearing to take place. Divalent cations such as Ca^{2+} and Mg^{2+} from the injected low-salinity brine control this process that results in ion exchange [17]. The requirements of this process are the presence of negatively charged surface on the rock, polar components on oil phase and divalent cations in the injection brine. MIE takes places during LSWF by removing organometallic complexes and polar compound from the clay surface and substituting them with noncomplex cations.

This theory was supported by experimental work carried out by Lager et al. [30] on core samples from North Slope composed of dead crude oil and initial connate water. Initially, the experiment was conducted at a temperature of 25°C, flooding the core with high-salinity brine followed by tertiary low-salinity flood, resulting in oil recovery of 42% OOIP for conventional high-salinity waterflooding and 48% OOIP for LSWF. A second experiment was run at 102°C, flooding the core samples with high-salinity water, and it resulted in an oil recovery of 35% OOIP. Divalent cations were removed from the cores by flushing it with brine containing high concentration of NaCl. The initial water saturation and oil condition were restored, and a high-salinity waterflood without the presence of divalent cations in the cores resulted in 48% OOIP, while no additional oil recovery was observed during LSWF.

From the experiment, it was concluded that the injection of low-salinity water into a sandstone reservoir in which mineral structure are not present will not result in incremental oil recovery. The findings also explained the reason why LSWF has no positive effect on acidized or fired sandstone as observed by Tang and Morrow [43] in their research. This was due to the absence of polar compounds that did not promote the interaction of clay minerals to release oil particles.

3.6.4 Limited release of mixed-wet particles

Limited release of mixed-wet particles is a combined mechanism of fine migrations proposed by Tang and Morrow [43] with DLVO theory. The name of DLVO theory originated upon the proposal of the theory by Derjaguin, Landau, Verwey and Overbeek. As explained by previous mechanisms, it is known that crude oil is originally bonded with clays which are attached on the pore surface. Due to the introduction of reduction in the salinity upon low-salinity water injection, the likelihood of these fines to be detached increases as the electrical double layer in the aqueous phase between is expanded. The migration and aggregation of stripped fines result in oil coalescing [38]. Oil recovery is enhanced due to the limited elimination of mixed-wet particles from the wall of pores because of local heterogeneous wetting conditions.

3.6.5 Wettability alteration

The alteration of the wettability of the reservoir is considered the primary mechanism of LSWF that results in incremental oil recovery. Previously explained mechanism such as fine migration, increase in pH and decrease in IFT, multi-component ion exchange and salt-in effect were related to the alteration of initial wettability of reservoir towards more water-wet. Suijkerbuijk et al. [34] related the changes in wettability in sandstone rocks with the presence of clays, composition of oil and high divalent cation concentration in formation water. The requirements for the positive effect of LSWF to take place were that the injection water also should contain divalent cations with the injection water salinity to be relatively lower than the salinity of formation water.

During low-salinity water injection into sandstone core, the mechanism of wettability alteration reported were similar to the process that occurred during alkaline flooding and surfactant flooding. The conducted experiment indicated an increase in pH value up to 10 that resulted in the generation of surfactant. This lowers the IFT between the water phase and the oil phase, thus increasing the water wettability promoting higher oil recovery [47]. Similar reaction mechanism occurred during the salting-in phenomenon with the decreasing salinity of injection brine.

The investigation of the effect of type of cation and its concentration in the injection water conducted by a researcher on the oil recovery of Berea sandstone concluded that wettability alteration was the main mechanism resulting in improvement in oil recovery [48]. The changes in the electrical charge upon LSWF in both the brine/oil and brine/rock interfaces to be more negative promote further stability of water film and result in water-wet state in the reservoir.

Moreover, several experiments conducted indicated that the contact angle between the oil and rock surfaces increases as the temperature and pressure increase and decrease with decreasing injection water salinity. As reported by Nasralla et al. [49], this observation was also supported by the increase in oil relative permeability end point and decrease in water relative permeability end point.

The low-salinity water flooding is an attractive eco-friendly and a promising technique for oil recovery in sandstone reservoirs in recent years. It can change the ion composition or brine salinity for improving oil recovery. However, the optimum conditions that improve oil recovery by low-salinity flooding are related to the understanding of fluid–rock interaction mechanisms. Low-salinity waterflooding might be effectively considered in special conditions for improving hydrocarbon recovery when the following factors are met: clay should be present in the sandstones, polar components (acidic and/or basic material) also should be present in crude oil, and formation water should contain divalent ions like Ca^{2+} [43, 50].

3.7 Aspects of modeling and simulation on low-salinity waterflooding

One of the earliest developments of the model to conduct studies on LSWF was accomplished by Jerauld et al. [32] through the modification of Buckley and Leveret conventional waterflooding model. In their model, the salinity of injection brine was made as a function of relative permeability and capillary pressure. The model built also includes the effects of secondary drainage water, relative permeability and hysteresis between imbibition and connate water. A similar LSWF model for sandstone and fractured media was presented by Wu and Bai [51] mathematically and numerically using MSFLOW general simulator. Results generated on the alteration of both models matched the experimental results, confirming the incremental oil recovery obtained via LSWF.

A semi-quantitative model developed by Sorbie [52] based on pore-scale theoretical considerations describes the multicomponent ion exchange mechanism of LSWF. This model was built with the purpose of demonstrating the effects of electrical double-layer expansion and polar organic species adsorption on the wettability alteration. Nevertheless, the predictions of the model were not convincing, and it was concluded that further experimental studies are required to validate it.

Omekeh [40] presented a mathematical model based on ion exchange and mineral dissolution and precipitation in LSWF. The model considered two-phase flow of oil and brine. From the research conducted, it was also demonstrated that the presence carbonate minerals may reduce the positive impact of LSWF in improving oil recovery. It was also proposed that cations are involved in an ion exchange process with the negatively charged clay surface and the release of cations from the surface of the rock increases relative permeability and mobility of oil. Desorption of divalent ions was suggested to be the main mechanism of LSWF. However, according to Suijkerbuijk [34], the proposed theory from the model contradicts with the experimental studies conducted on the mechanism of LSWF. Adsorption of divalent ions on the clay minerals was reported to be the process that alters wettability resulting in LSWF benefits.

Recently, a systematic study of LSWF mechanism and its potential in improving oil recovery was presented by Dang et al. [52] using a mechanistic model that was developed using Computer Modeling Group's GEM™ reservoir simulator validated against PHREEQC geochemistry software and few other experimental coreflooding tests. In this model, the role played by clay was captured in investigating the geological effects in the process of LSWF, and the field-scale benefits of LSWF in both secondary and tertiary injection modes were proven. Changes in wettability condition due to ion exchange and clay effects were proposed to be the primary mechanism of LSWF, and nominal optimization of the process was presented in this research. **Table 2** summarizes past modeling and simulation studies that were conducted to understand the process of LSWF.

Author	Software used	Research scope
[32]	Buckley and Leveret model	Effect of injection water salinity on relative permeability and capillary pressure
[47]	PHREEQC geochemical software	Changes in pH of reservoir during LSWF
[51]	Buckley and Leveret model	Relationship between injection salinity concentration and wettability alteration
[52]	PHREEQC geochemical software	Description of the multicomponent ion exchange process at the pore scale
[40]	Mathematical and salt reaction model	Modeling of ion exchange and mineral solubility in LSWF
[53]	IPHREEQC geochemical module coupled with UTCHEM chemical flooding reservoir simulator	Multicomponent ion exchange mechanism and effect on LSWF
[54]	PHREEQC geochemistry software coupled with CMG's GEM	Investigation on effect of clays and ion exchange process on LSWF and process optimization

Table 2.
Summary of previous modeling and simulation studies related to low-salinity waterflooding.

4. Conclusion

The evaluation on the prominent ideas and aspects of LSWF were presented in this paper mainly focusing on sandstone reservoir. Different characteristics of LSWF have been thoroughly reviewed including the industrial application, field studies, mechanisms, laboratory and modeling works that have been conducted. Based on previous reports, LSWF has a beneficial effect on oil recovery in both laboratory and field-scale studies. The mechanisms that resulted in incremental oil recovery compared to standard high-salinity waterflooding that have been proposed by several researches over the years are fine migration, increase in pH and reduction in IFT, multicomponent ion exchange, limited release of mixed-wet particles and wettability alteration. Some of these mechanisms are related to each other with the main process being wettability alteration.

It can be deduced that there is no general agreement regarding which mechanism results in incremental oil recovery, and these mechanisms work under a specific condition during low-salinity water injection. The magnitude of incremental oil recovery obtained via LSWF is highly dependent on the reservoir condition as the working mechanism directly relates to the specifics of the reservoir because the wettability can be changed from oil-wet to water-wet or from water-wet to mixed-wet. Although in either way oil recovery factor could be improved, the magnitude of oil recovery may vastly vary. Another fact is that generally LSWF is used together with chemical flooding. LSWF is largely environmentally friendly compared to chemical methods and has higher oil recovery benefits than conventional waterflooding method. However, considering the incremental oil recovery from chemical EOR projects, the incremental oil recovery from LSWF alone should not be too high.

Author details

Hisham Ben Mahmud[1]*, Shattia Arumugam[2] and Walid Mahmud[3]

1 Curtin University, Malaysia

2 Weatherford (Malaysia) Sdn Bhd

3 University of Tripoli, Libya

*Address all correspondence to: hisham@curtin.edu.my

References

[1] Santa M, Alvarez-Jürgenson G, Busch S, Birnbrich P, Spindler C, Brodt G. Sustainable surfactants in enhanced oil recovery. In: SPE Enhanced Oil Recovery Conference. Kuala Lumpur, Malaysia: Society of Petroleum Engineers; 19-21 July 2011. Available from: https://doi-org.dbgw.lis.curtin.edu.au/10.2118/145039-MS

[2] Sajjad FM. Smart Water EOR Effects in Preserved Sandstone Reservoir Cores, Comparison Between Sea Water and Low Salinity Brines at 136°C. 2015. Retrieved from https://brage.bibsys.no/xmlui/handle/11250/301284

[3] Dang C, Nghiem L, Modelling C, Nguyen N, Chen Z, Yang C, et al. A comprehensive evaluation of alkaline surfactant polymer flooding and hybrid process for enhanced oil recovery state-of-the-art development of ASP flooding. In: SPE Annual Technical Conference and Exhibition. San Antonio, Texas, USA: Society of Petroleum Engineers; 9-11 October 2017. Available from: https://doi-org.dbgw.lis.curtin.edu.au/10.2118/187132-MS

[4] Riley RA, Harper JA, Harrison WB III, Barnes D, Nuttall B, Avary KL, et al. Evaluation of CO_2—Enhanced oil recovery and sequestration opportunities in oil and gas fields in the MRCSP region. MRCSP Phase II Topical Report. DOE-NETL Cooperative Agreement DE-FC26-05NT42589. October 2005-2010

[5] Sheng JJ. Critical review of low-salinity waterflooding. Journal of Petroleum Science and Engineering. 2014;**120**:216-224. DOI: 10.1016/j.petrol.2014.05.026

[6] Aladasani A, Bai B, Wu YS, Salehi S. Studying low-salinity waterflooding recovery effects in sandstone reservoirs. Journal of Petroleum Science and Engineering. 2014;**120**:39-51. DOI: 10.1016/j.petrol.2014.03.008

[7] Yrigoyen A, Saputelli L, Chacon A, Chegin J. SPE-187300-MS identifying cost-effective waterflooding optimization opportunities in mature reservoirs from data driven analytics. In: SPE Annual Technical Conference and Exhibition. San Antonio, Texas, USA: Society of Petroleum Engineers; 9-11 October 2017. Available from: https://doi-org.dbgw.lis.curtin.edu.au/10.2118/187300-MS

[8] Al-shalabi EW, Sepehrnoori K, Pope G. Mechanistic modeling of oil recovery due to low salinity water injection in oil reservoirs. In: SPE Middle East Oil & Gas Show and Conference. Manama, Bahrain: Society of Petroleum Engineers; 8-11 March 2015. Available from: https://doi-org.dbgw.lis.curtin.edu.au/10.2118/172770-MS

[9] Melo MA, Almeida AR. SPE-184956-MS determining the sweep efficiency of waterflooding using tracers. In: SPE Latin America and Caribbean Mature Fields Symposium. Salvador, Bahia, Brazil: Society of Petroleum Engineers; 15-16 March 2017. Available from: https://doi-org.dbgw.lis.curtin.edu.au/10.2118/184956-MS

[10] Beliveau D. SPE 113132 waterflooding viscous oil reservoirs. Pdf. SPE Reservoir Evaluation & Engineering. Society of Petroleum Engineers; October 2009;**12**(5):4-6. Available from: https://doi-org.dbgw.lis.curtin.edu.au/10.2118/113132-PA

[11] Erzuah S, Fjelde I, Omekeh AV. Wettability estimation by surface complexation simulations. In: SPE Europec featured at 79th EAGE Conference and Exhibition. Paris, France: Society of Petroleum Engineers; 12-15 June 2017. Available from: https://doi-org.dbgw.lis.curtin.edu.au/10.2118/185767-MS

[12] Hirasaki GJ. Wettability: Fundamentals and surface forces. SPE Formation Evaluation. 1991;**6**(02): 217-226. DOI: 10.2118/17367-PA

[13] Anderson W. Wettability literature survey—Part 2: Wettability measurement. Journal of Petroleum Technology. 1986;**38**(11):1246-1262. DOI: 10.2118/13933-PA

[14] Alnili F, Al-Yaseri A, Roshan H, Rahman T, Verall M, Lebedev M, et al. Carbon dioxide/brine wettability of porous sandstone versus solid quartz: An experimental and theoretical investigation. Journal of Colloid and Interface Science. 2018;**524**:188-194. DOI: 10.1016/j.jcis.2018.04.029

[15] Agada S, Geiger S, Doster F. Wettability, hysteresis and fracture-matrix interaction during CO_2 EOR and storage in fractured carbonate reservoirs. International Journal of Greenhouse Gas Control. 2016;**46**:57-75. DOI: 10.1016/j.ijggc.2015.12.035

[16] Askarinezhad R, Hatzignatiou DG, Stavland A. Disproportionate permeability reduction of water-soluble silicate gelants—Importance of formation wettability. In: SPE Improved Oil Recovery Conference. Tulsa, Oklahoma, USA: Society of Petroleum Engineers; 11-13 April 2016

[17] AlRatrout A, Raeini AQ, Bijeljic B, Blunt MJ. Automatic measurement of contact angle in pore-space images. Advances in Water Resources. 2017;**109**:158-169. DOI: 10.1016/j.advwatres.2017.07.018

[18] Mugele F, Siretanu I, Kumar N, Bera B, Wang L, de Ruiter R, et al. Insights from ion adsorption and contact-angle alteration at mineral surfaces for low-salinity waterflooding. SPE Journal. 2016;**21**(4):1-10. Available from: http://www.onepetro.org/doi/10.2118/169143-PA

[19] Etemadi A, Khodapanah E, Tabatabaei-Nejad SA. Modelling low-salinity waterflooding: Effect of divalent cations and capillary pressure. Journal of Petroleum Science and Engineering. 2017;**149**:1-8

[20] Anderson WG. Wettability literature survey—Part 1: Rock/oil/brine interactions and the effects of core handling on wettability. Journal of Petroleum Technology. 1986;**38**(10):1125-1144

[21] Buckley JS, Bousseau C, Liu Y. Wetting alteration by brine and crude oil: From contact angles to cores. In: SPE Annual Technical Conference and Exhibition. SPE Journal. Dallas: Society of Petroleum Engineers; 1995. SPE 30765

[22] Tang GQ, Morrow NR. Effect of temperature, salinity and oil composition on wetting behavior and oil recovery by waterflooding. In: SPE Annual Technical Conference and Exhibition, Denver; 1996. SPE 36680

[23] Cuiec LE. Rock/crude-oil interactions and wettability: An attempt to understand their interrelation. In: SPE 59th Annual Technical Conference and Exhibition, Houston, Texas; 1984. SPE 13211

[24] Clementz DM. Alteration of rock properties by adsorption of petroleum heavy ends: Implications for enhanced oil recovery. In: SPE/DOE 3rd Joint Symposium on Enhanced Oil Recovery, Tulsa, Oklahoma; 1982. SPE/DOE 10683.

[25] Brown CE, Neustadter EL. The wettability of oil/water/silica systems with reference to oil recovery. Journal of Canadian Petroleum Technology. 1980;**19**(3):100-110

[26] Physics of Complex Fluids, University of Twente. 2017. Available from: https://www.utwente.nl/en/

tnw/pcf/science/current_projects/Enhanced_Oil_Recovery/

[27] Hughes RV, Pfister RJ. Advantages of brines in secondary recovery of petroleum by water-flooding. Transactions of the American Institute of Mining, Metallurgical, and Petroleum Engineers. 1947;**170**:187-201

[28] Reiter PK. A water-sensitive sandstone flood using low salinity water (master of science thesis). USA: University of Oklahoma; 1961

[29] Bernard GG. Effect of floodwater salinity on recovery of oil from cores containing clays. In: SPE California Regional Meeting. Society of Petroleum Engineers; 1967. pp. 1-8

[30] Lager A, Webb KJ, Black CJJ, Singleton M, Sorbie KS. Low salinity oil recovery—An experimental Investigation1. Petrophysics. 2008;**49**(1):28-35

[31] Zhang Y, Xie X, Morrow NR. Waterflood performance by injection of brine with different salinity for reservoir cores. In: Proceedings of the SPE Annual Technical Conference and Exhibition. Vol. 2. 2007. pp. 1217-1228. Anaheim, California, USA: Society of Petroleum Engineers; 11-14 November 2007. Available from: http://www.scopus.com/inward/record.url?eid=2s2.040449142223&partnerID=tZOtx3y1

[32] Jerauld G, Webb K, Lin C-Y, Seccombe J. Modeling low-salinity waterflooding. SPE Reservoir Evaluation and Engineering. 2008;**11**(6):24-27

[33] Aladasani A, Bai B, Wu Y. Investigating low-salinity waterflooding recovery mechanisms in sandstone reservoirs. In: Eighteenth SPE Improved Oil Recovery Symposium. Tulsa, Oklahoma, USA: Society of Petroleum Engineers; 14-18 April 2012. Available from: ///users/shattia/downloads/mechanism/SPE-172183-MS.pdf

[34] Suijkerbuijk BM, Sorop TG, Parker AR, Masalmeh SK, Chmuzh IV, Karpan VM, et al. Low salinity waterflooding at west-salym: Laboratory experiments and field forecasts. In: SPE Improved Oil Recovery Symposium. Tulsa, Oklahoma, USA: Society of Petroleum Engineers; 12-16 April 2014. Available from: http: //www.onepetro.org/doi/10.2118/169102-MS

[35] Zhang Y, Morrow NR. Comparison of secondary and tertiary recovery with change in injection brine composition for crude-oil/sandstone combinations. In: SPE/DOE Symposium on Improved Oil Recovery. Tulsa, Oklahoma, USA: Society of Petroleum Engineers; 22-26 April 2006. Available from: http://www.onepetro.org/doi/10.2118/99757-MS

[36] Patil SB, Dandekar AY, Patil S, Khataniar S. Low salinity brine injection for EOR on Alaska north slope (ANS). In: International Petroleum Technology Conference. Kuala Lumpur, Malaysia: International Petroleum Technology Conference; 3-5 December 2008. DOI: 10.2523/IPTC-12004-MS

[37] Agbalaka CC, Dandekar AY, Patil SL, et al. Coreflooding studies to evaluate the impact of salinity and wettability on oil recovery efficiency. Transport in Porous Media. 2009;**76**(1):77-94

[38] Rivet S, Lake LW, Pope GA. A coreflood investigation of low-salinity enhanced oil recovery. In: SPE's Annual Caspian Technical Conference and Exhibition. 2010. pp. 19-22. Available from: http: //www.onepetro.org/doi/10.2118/134297-MS

[39] Zhang C, Gamage RP, Perera MSA, Zhao J. Characteristics of clay-abundant shale formations: Use of CO_2 for production enhancement. Energies. 2017;**10**(11):10111887

[40] Omekeh AV, Friis HA, Fjelde I, Evje S. Modeling of ion-exchange and

solubility in low salinity water flooding. In: SPE Improve Oil Recover Symposium. 2012. Available from: http://www.onepetro.org/doi/10.2118/154144-MS

[41] Sheng JJ. Critical review of low-salinity waterflooding. Journal of Petroleum Science and Engineering. 2014;**120**:216-224. DOI: 10.1016/j.petrol.2014.05.026

[42] Jackson MD, Vinogradov J, Hamon G, Chamerois M. Evidence, mechanisms and improved understanding of controlled salinity waterflooding part 1: Sandstones. Fuel. 2016;**185**:772-793. DOI: 10.1016/j.fuel.2016.07.075

[43] Tang GQ, Morrow NR. Influence of brine composition and fines migration on crude oil/brine/rock interactions and oil recovery. Journal of Petroleum Science and Engineering. 1999;**24**(2-4):99-111. DOI: 10.1016/S0920-4105(99)00034-0

[44] Zeinijahromi A, Bedrikovetsky P. Physics mechanisms of enhanced recovery by fines-migration-assisted waterflooding (laboratory study). In: SPE European Formation Damage Conference & Exhibition. Noordwijk, The Netherlands: Society of Petroleum Engineers; 5-7 June 2013

[45] Mcguire PLL, Chatham JRR, Paskvan FKK, Sommer DMM, Carini FHH, Exploration BP. Low salinity oil recovery: An exciting new EOR opportunity for Alaska's North Slope. In: SPE Western Regional Meeting. 2005. pp. 1-15. Available from: https://www.onepetro.org/conference-paper/SPE-93903-MS

[46] Austad T. Water-based EOR in carbonates and sandstones: New chemical understanding of the EOR potential using "smart water". In: Enhanced Oil Recovery Field Case Studies. Elsevier Inc. Gulf Professional Publishing; 2013. pp. 301-335. DOI: 10.1016/B978-0-12-386545-8.00013-0

[47] Rueslatten HG, Hjelmeland O, Selle OM. Wettability of reservoir rocks and the influence of organo-metallic compounds. North Sea Oil and Gas Reservoir. 1994;**3**:317-324

[48] Dang CTQ, Nghiem L, Modelling C, Nguyen NTB, Zhangxin C. New insights into the critical role of geology in modeling and prediction of low salinity waterflooding. Europec, Madrid, Spain: Society of Petroleum Engineers; 1-4 June 2015. Available from: https://doi-org.dbgw.lis.curtin.edu.au/10.2118/174294-MS

[49] Nasralla RA, Nasr-El-Din HA. Double-layer expansion: Is it a primary mechanism of improved oil recovery by low-salinity waterflooding? SPE Reservoir Evaluation and Engineering. 2014;**17**(01):49-59. Available from: http://www.onepetro.org/doi/10.2118/154334-PA

[50] Lager A, Webb K, Black C. Impact of brine chemistry on oil recovery. In: IOR 2007-14th European Symposium on Improved Oil Recovery. Emerging Technologies for Hydrocarbon Recovery. EAGE; 22 April 2007

[51] Wu YS, Bai B. Efficient simulation for low salinity waterflooding in porous and fractured reservoirs. In: SPE Reservoir Simulation Symposium. The Woodlands, Texas: Society of Petroleum Engineers; 2-4 February 2009. Available from: http://www.onepetro.org/doi/10.2118/118830-MS

[52] Sorbie KS, Collins I. A proposed pore-scale mechanism for how low salinity Waterflooding works. In: SPE Improve Oil Recover Symposium. 2010. pp. 1-18. Available from: http://www.onepetro.org/doi/10.2118/129833-MS

[53] Korrani AKN, Sepehrnoori K, Delshad M. A novel mechanistic approach for modeling low salinity

water injection. In: Paper SPE 166523 presented at the SPE ATCE; New Orleans, Louisiana, USA; 2013

[54] Dang CTQ, Nghiem L, Nguyen NTB, Chen Z. Practical concerns and principle guidelines for screening, implementation, design, and optimization of low salinity waterflooding. In: SPE Western Regional Meeting. Garden Grove, California, USA: Society of Petroleum Engineers; 27-30 April 2015

Chapter 6

Development and Application of Chemical EOR Technologies in China Offshore Oil Fields

Jian Zhang, Fengjiu Zhang, Xiaodong Kang and Baozhen Li

Abstract

At present, polymer flooding as the most effective chemical EOR technique is widely used in onshore oil fields in the world. Also, it has been successfully applied in China offshore oil fields as a major EOR technology. CNOOC has preliminarily established a chemical flooding (polymer, polymer-surfactant, weak gel, etc.) technology system including high-efficiency chemical flooding agents, platform injection facilities, and produced liquid treatment technology. Since 2003, pilot tests and field applications were carried out in S, L, and JW oil fields, and predicted oil increment and good economic benefits have been achieved, which proved that offshore chemical EOR technology is feasible and economical. It has explored a new road for increasing the recovery of offshore oil fields and provided a solid technical guarantee for their economic and efficient development.

Keywords: offshore oil field, chemical flooding, EOR, review

1. Introduction

By the end of December 2017, China's offshore oil production accounted for 22.3% of the national oil production [1–4]. The average offshore field water cut of oil fields was 86.6%, with relatively low oil recovery rate, only 21.1%. Bohai oil field is the largest offshore oil field in China. In 2010, its oil and gas output reached 30 million square meters, accounting for about 15% of the national total, which played an important role in stabilizing the energy supply in eastern China. Heavy oil accounts for 85% of the oil reserves in Bohai oil field, and the main development bottlenecks of these heavy oil fields are high crude oil viscosity, strong reservoir heterogeneity, and poor water drive efficiency. In particular, the viscosity of underground crude oil in some heavy oil reservoirs reaches 200–700 mPa s, and their recovery factor from traditional water flooding could only reach about 12% [5–9]. It could be seen that the research and application of enhanced oil recovery technology in offshore heavy oil reservoirs are of great significance for the stable production and economic efficient development of Bohai oil field.

2. The technical challenges of offshore chemical flooding

Chemical flooding has been successfully carried out in onshore oil fields in China for nearly 30 years. The implementation of traditional polymer flooding technology requires a large amount of freshwater supply, large injection allocation equipment, good reservoir conditions, and loose environmental requirements. It is difficult to meet these requirements in offshore oil fields. To realize the field application of polymer flooding technology, there are several challenges to overcome to conduct offshore chemical EOR projects:

1. Lack of fresh water on offshore platforms. It is almost impossible to use fresh water to prepare polymer for offshore polymer flooding. Only seawater (salinity 32,000–35,000 mg/L) or formation water could be considered.

2. Narrow space of offshore platform. The equipment for polymer solution preparing and injection is required to be compact, flexible, and more efficient. At the same time, the polymer is required to have excellent and efficient viscosity properties, including temperature and salt resistance and instant solubility characteristic.

3. The reservoir conditions are complex. For example, most Bohai oil fields have many high-permeability heterogeneous layers, high-salinity formation water, and high-viscosity crude oil.

4. Large well space. Most offshore oil fields were developed with large well space (350–600 m) and anti-nine-spot well pattern.

5. Environmental requirements for offshore oil field development are high. It is difficult for produced fluid from polymer flooding to meet standards in the last treatment on the platform, which will have a great impact on water reinjection.

6. There is a big difference between offshore oil field and onshore ones in development model and polymer injection timing, so it is necessary to establish evaluation methods and standard for offshore polymer flooding performance [5–9].

3. Research and development of offshore chemical EOR techniques

It could be seen that the research and application of enhanced oil recovery technology in offshore heavy oil reservoirs are of great significance for stale production and economic and efficient development of Bohai oil field. Based on about 20 years of research offshore chemical EOR techniques, a series of resolutions and techniques were developed including high-efficiency driving agents, platform polymer injection distribution technology, produced fluid treatment technology, and performance evaluation method for early-stage polymer flooding. These techniques had been carried out in S, L, and JW oil fields. The water cut was controlled at a low level, and the incremental oil was obviously enhanced.

3.1 Development and application of novel salt-resistant polymers

Bohai oil field has the characteristics of high crude oil viscosity, large well spacing, high water salinity, and limited platform space. Thus, the suitable polymer agent should have high-viscosity property, good salt tolerance, aging stability, and fast-dissolution ability. In order to expand polymer injection scale, two kinds of salt-resistant polymer were investigated and applied in Bohai Bay.

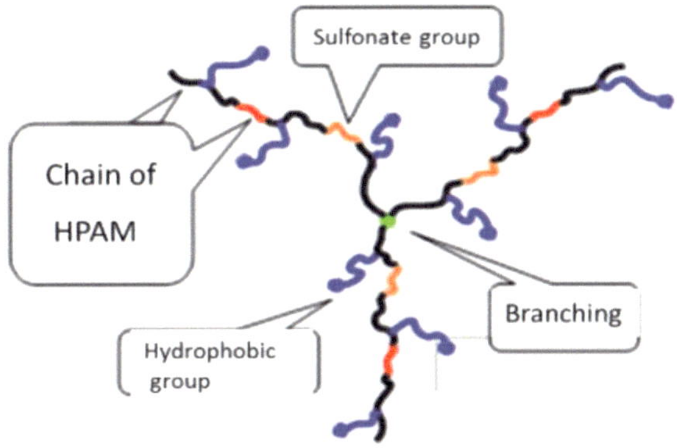

Figure 1.
Schematic of hydrophobic associated polymer.

3.1.1 Hydrophobic associated polymer

With reference to the research of hydrophobic acrylamide polymers and the literature on the synthesis of ultrahigh-molecular-weight hydrolyzed polyacrylamide, a new kind of polymer was designed with branched structure and hydrophobic associative units with micro-blocks (**Figure 1**). Meanwhile, a high-molecular-weight hydrophobic associative polymer industrial synthesis technology and polymer fast-dissolution technology was developed which could meet the requirements of large-scale application of high-efficiency offshore EOR technology [10, 11]. Under above conditions, the viscosity of hydrophobic associated polymer is 3–5 times that of ultrahigh-molecular-weight HPAM, the hard water resistance is increased by three times, the shear and aging retention rate are increased by 1.6 times, the residual resistance coefficient are increased by 2–3 times, and the dissolution time is shortened to 40 min.

Hydrophobic associated polymer has been applied in Bohai oil field since September 2003 and has achieved significant oil and precipitation effects [7–10].

3.1.2 Structural composite polymer

The performance of polymer depends on the molecular structure. Structural composite polymer is a kind of polymer with high sterically hindered side group on its skeleton, which exhibits relative rigidity in the solution, and the polymer molecule is strengthened under relatively harsh conditions by increasing the kinetic radius. Also a functional monomer which adjusts the amphiphilic nature of the polymer molecule is introduced into the molecular structure of the polymer to improve the solubility of the polymer. Structurally, polymer molecules contain diversified and multifunctional monomers, hence the name structural composite polymers (**Figure 2**). By adjusting the type and proportion of functional monomers, the salt and temperature resistance ability of polymers was improved. The polymer at lower concentrations could significantly reduce the viscosity of the heavy oil and significantly improves the fluidity of the heavy oil [12–14]. Above polymer had been injected in several pilot wells in Bohai oil field since December 2018.

3.2 Chemical flooding injection system on offshore platform

In order to promote the application of polymer flooding technology in offshore oil fields, it was necessary to combine the special conditions of offshore platforms to break through the bottleneck limitation of the huge polymer dispensing system

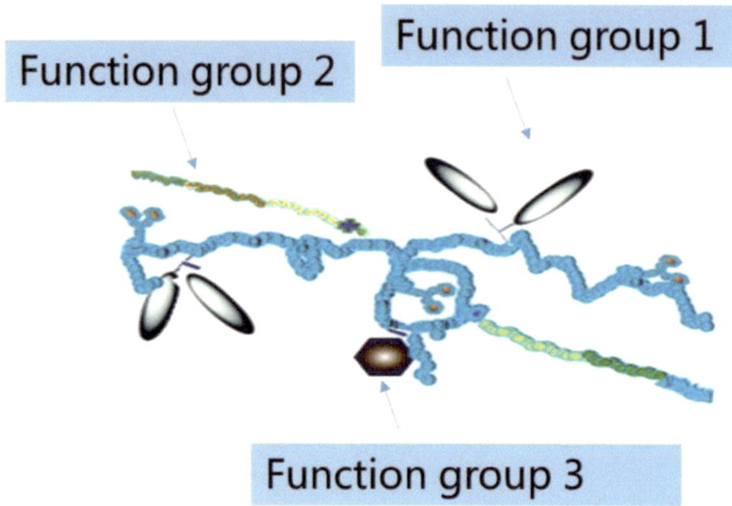

Figure 2.
Schematic of structural composite polymer.

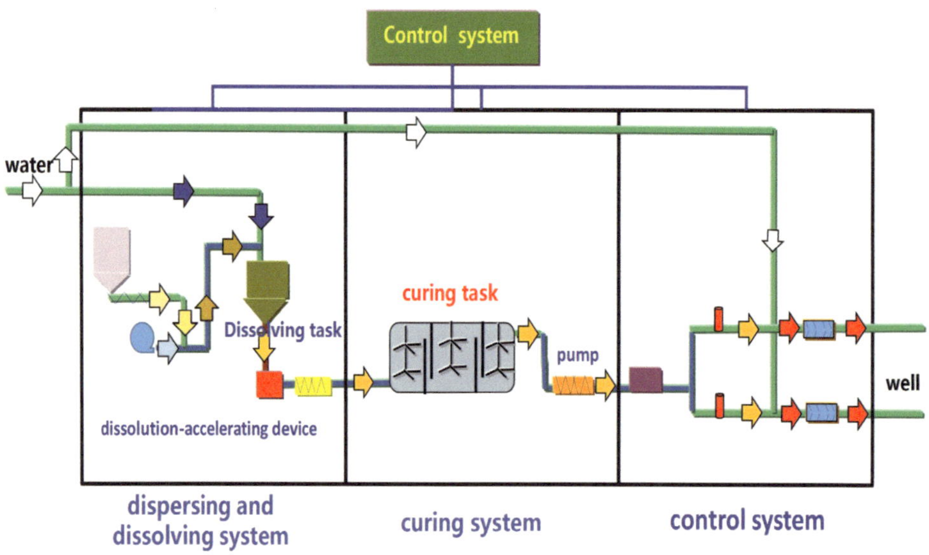

Figure 3.
The process of offshore platform injection system.

and achieve miniaturization, high efficiency, and modularization of the polymer injection system [15, 16]. The offshore polymer dispensing equipment included four systems: the dispersing and dissolving system, the curing system, the high-pressure injection system, and the control system (**Figure 3**). Each system was independent and convenient to install according to the site conditions.

Dispersing and dissolving system was the core of the rapid dissolving device. According to their viscoelastic characteristics, swelling polymer micelles were stretched during going through the gaps of the stator and rotor by using moderate stretching method. More water molecules permeate into the inside of the polymer micelle. At the same time, polymer molecules at the micelle's surface are stretching down into the water. Curing system achieved the full dissolution and ripening of the polymer solution. The size was dependent on the injection volume and curing time of the polymer solution. In order to ensure fully dissolution, the interior of the curing tank was optimized. The curing tank was separated into the inlet chamber, ripening chamber, and outlet chamber through a choke plate [15].

High-pressure injection system was applied to achieve the flow distribution and pressurization of the polymer mother liquor. The polymer mother liquor is distributed by a flow distributor for each well and then pressurized by a high-pressure injection pump.

Control system was applied to achieve automatic control of the entire polymer dispensing system, automatic data acquisition, and variable frequency control equipment for power equipment.

The online polymer dissolving technology was combined with technologies of polymer powder transmission by positive pressure gas, multistage polymer dispersion and stretching, graded maturation, and high efficient mixing, as the polymer dissolving time shortened to 40 min with high-viscosity retention rate (91%). Under same injection amount, area was saved by 21%, and weight was reduced by 37%. The improved facility has operated steadily in Bohai oil field for 2 and a half years.

3.3 Reservoir dynamic analysis method for offshore chemical flooding fields

The reservoir dynamic analysis method for polymer flooding in onshore oil fields is mature where the polymer injection is started when water cut is over 80%. However, in some offshore field, polymer injection was conducted at very low water cut (even <20% in some case) for the limited life of platform [6–8, 16]. The traditional polymer flooding evaluation method is not applicable in offshore early-stage polymer flooding projects for their short water flooding period, and the effectiveness of injection and the calculation method of incremental oil are not applicable. Based on the power law fluid's stable flow in pore media, considering non-Newtonian and multilayer heterogeneity of polymer solution, a Hall curve evaluation method of early polymer flooding is established. The principle is to use pressure and injection data to draw improved Hall curve of polymer flooding in linear form.

From **Figure 4**, it could be seen that, unlike the traditional Hall curve of polymer flooding, the slopes and intercepts of the improved Hall curve could be used to directly calculate the fluidity and resistance coefficients during polymer flooding, thus solving the reservoir dynamic analysis problem in early-stage polymer injection projects.

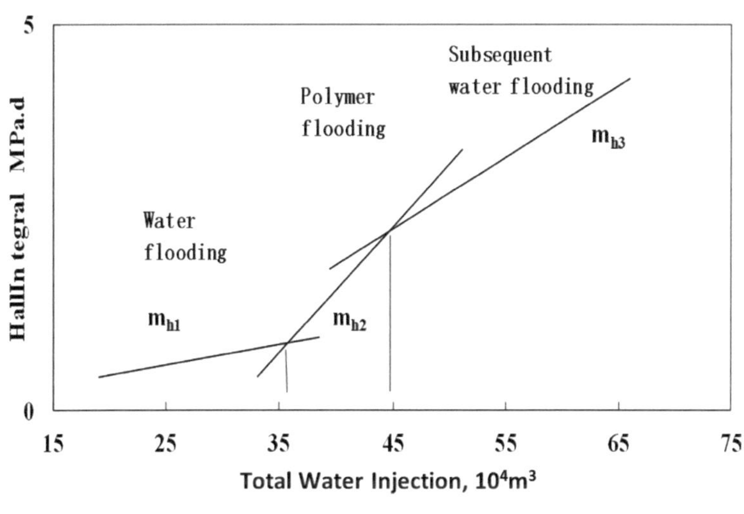

Figure 4.
Schematic diagram of calculation for resistance coefficient and residual resistance coefficient.

3.4 Treatment technology of chemical flooding produced fluid in offshore platform

Partially hydrolyzed polyacrylamides with high molecular weight (HPAM), after undergoing a series of chemical, mechanical, and thermal degradations, are back produced, making the oil-water interface highly electronegative and forming a strong electrical double layer. The repulsive force between oil droplets affects their aggregation and coalescence, and at the same time, the adsorbed polymers also increase the strength and viscoelasticity of the oil-water interface film. As a result, the emulsified oil droplets are of small particle size and strong stability. Therefore, the oil-water separation of produced fluid from polymer flooding is more difficult than that from water flooding. Hence, devices with much longer residence time and several times larger size are required for produced water treatment [17, 18].

However, platform space is limited for offshore oil fields; the abovementioned practice cannot be implemented. At present, the treatment process of polymer-containing produced fluid in Bohai oil fields is a so-called three-stage treatment process similar to that of water flooding. For example, the comprehensive water content of the produced fluid is 78% in Bohai A oil fields, and the average polymer concentration is 150 mg/L. The oil treatment system includes a one-stage three-phase separator, a two-stage thermal settling separator, and a conventional electric dehydrator (**Figure 5**). The wastewater treatment system mainly includes a tilted plate separator, a gas flotation unit (GFU), a walnut shell filter, and a residue conversion system. The abovementioned process represents the traditional process of produced fluid treatment for water flooding oil fields. For different polymer flooding oilfields, the original water flooding process have different complexity and methods for adaptive modification for polymer flooding projects.

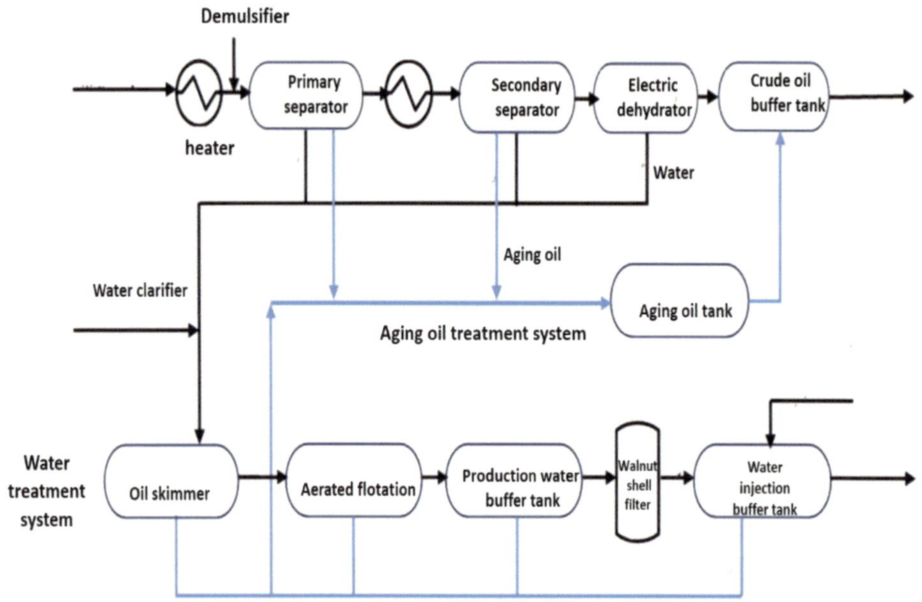

Figure 5.
Treatment process of produced fluid in Bohai A oil field.

4. Field application of offshore chemical EOR technology

In order to improve the development effect of offshore oil field during the effective period of platform, offshore chemical flooding was carried out in three Bohai oil fields since 2003 (**Table 1**), and good economic benefits have been achieved [6, 7, 19–25].

4.1 Polymer flooding in S field

S oil field is located at the coast of Liaodong, Bohai Bay, with stable distribution and good connectivity. Its reservoir rocks are feldspathic quartz sand composed mainly of fine sands. The crude oil in S oil field is highly viscous, varying from 13 to 380 mPa s, with an average of 70 mPa s at reservoir conditions. It was put into operation since 1993 with inversed 9-point well pattern. From the very beginning, the enhanced oil recovery technologies were taken into consideration, and polymer flooding showed great potential among these technologies. Since September 2003, the first single-well polymer injection pilot in China offshore field was successfully carried out in S field [19–21]. In October 2005, the well group polymer injection was carried out in AJ platform in S oil field, which consisted of four injectors and one central producer and seven peripheral producers. After 3 years of well group test, the expected incremental oil was obtained. Based on the well group polymer injection test result in S field, the expanded polymer injection project was implemented in mid-2008. Up to 2014, 24 wells were successively converted to polymer injectors, and significant incremental oil was achieved (**Figure 6**). By the end of 2017, 4.75 million cubic meters of cumulative oil were achieved, increasing recovery rate by 5.2%.

4.2 Chemical flooding in L field

L oil field is located in Bohai Bay, characterized by huge thickness, high permeability, severe heterogeneity, high crude oil density (0.947 g/cm^3), and medium oil viscosity (7.2–19.4 cp). L oil field was put into production in January of 2005 and started to inject water in September of 2005. Based on well understanding of the mechanism and effect of the early polymer flooding, single-well polymer

Indexes		S	L	JW	Daqing
Reservior Oil Viscosity (mPa. s)		70	16	17	8
		13.3-442.2	13.9-19.4	6.1-26.5	7--10
Well Pattern		Inversed 9 point pattern			5-spot pattern
Well Spacing (m)		350-590	350-400	400	106-200
Average Thickness (m)		32.4	45.2	19.3	11-21
Injected water (mg/L)	Salinity	9600	2873	3000	4600
	Divalent cations	810	260	34	20
Polymer Injection Time	Water Cut (%)	68.0	8.5	79.0	83-95
	Developing Stage	Mid-Mid high	Low	Mid-Mid high	High-Ultra high
Characteristics of Pilot Fields		High oil Viscosity; High levels of divalent cations	Block Reservoir; Polymer injeciton at low Water Cut	The most similar to onshore oilfields	Light oil; Good water; thin Layers; uncompartmentalized oilfield

Table 1.
Comparison of the three offshore chemical flooding fields with Daqing Field.

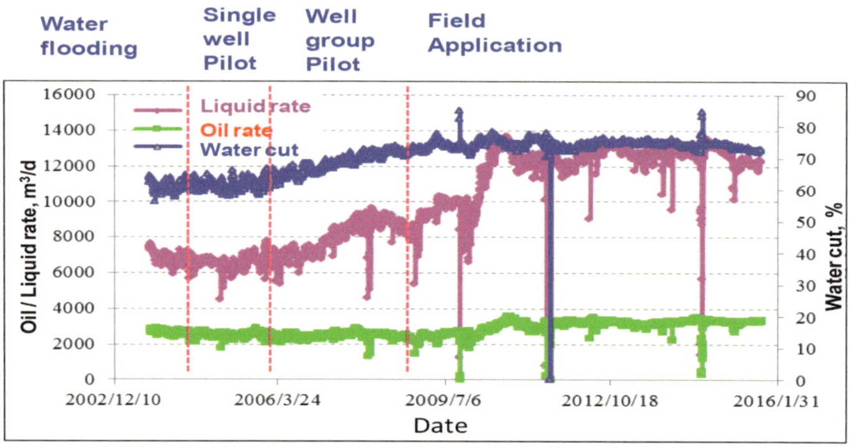

Figure 6.
Field production curves in S field.

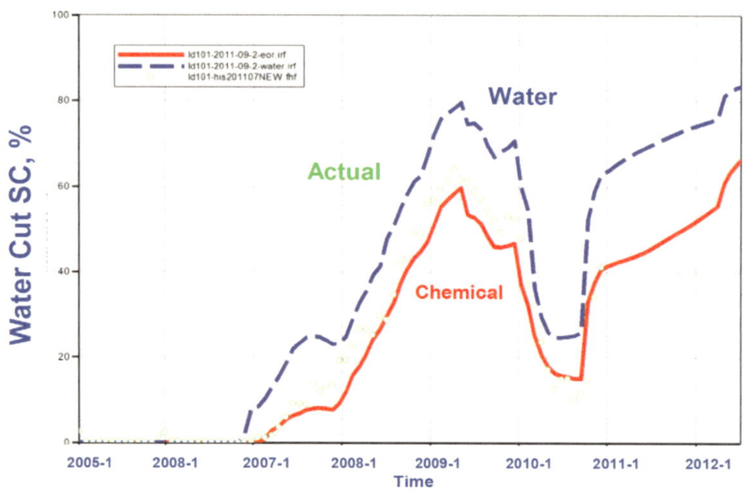

Figure 7.
Water cut curves under water flooding, chemical flooding, and actual situation.

injection pilot test was carried out since 2006 when the water cut in the pattern was lower than 10% [22–24]. After that, another five water injectors were converted to polymer injectors from 2007 to 2009. For the early-stage polymer flooding, the characteristics of the responses on producers were different from the case in which polymer flooding was conducted during high water cut stage. The water production of the producers continued to rise up after polymer flooding, but the simulation research showed that the water cut increasing rate was lower than the rate during water flooding. Of course, water cut drop was observed in some wells (**Figure 7**). By December 2014, the total incremental oil by polymer flooding was about 754,650 m^3, and the stage oil recovery was enhanced by 3.0%. The polymer flooding is still effective, and more incremental oil will be obtained later.

4.3 Chemical flooding in JW field

The water depth of W oil field is from 6.5 to 10.5 m; the reservoir depth is 1600–1800 m with banded fault anticline structural form and 25 m average effective thickness. This reservoir has normal temperature and pressure system containing oil with

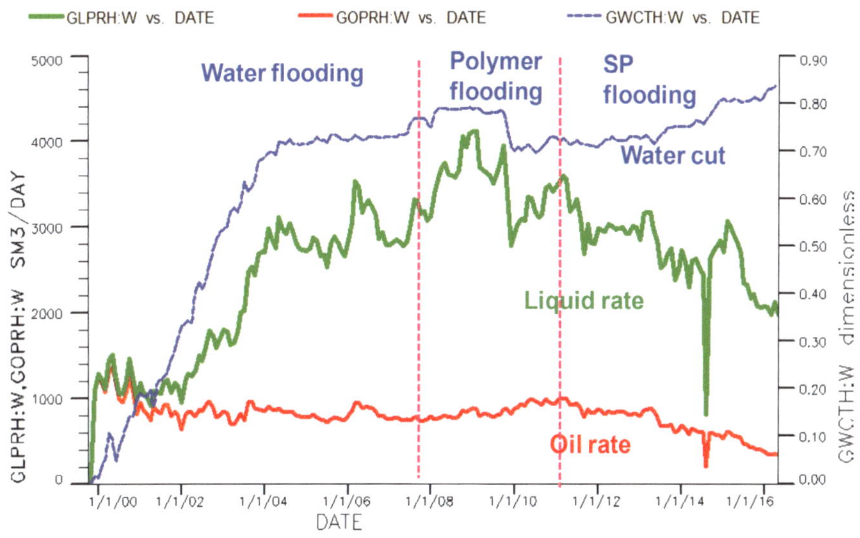

Figure 8.
Field production curves in JW field.

underground density of 0.87 g/cm^3 and reservoir viscosity of 16–26 mPa s. The formation porosity is from 22 to 36% and permeability from 0.01 to 5 μm^2. There are many sub-reservoirs in the upper part of this oil field, with large permeability difference and serious heterogeneity [25, 26]. W oil field was put into production since 1999 with anti-nine-spot well pattern and 350–400 m well space. Water flooding began in December 2000. This oil field was considered as the best candidate for polymer flooding which has the lowest water salinity and medium oil viscosity in Bohai Bay. In order to improve the efficiency of water flooding and reduce the decline rate of production, eight polymer injectors were gradually implemented at field water cut of 79% since 2007. The injection rate was 0.045 PV/year, and the polymer injection concentration was 1200 mg/L. After polymer flooding, the characteristics of polymer flooding such as decreasing water cut drop, increasing oil production, and injection profile improvement were observed gradually. In order to improve the performance of chemical flooding, eight polymer injectors were transferred into polymer-surfactant injectors since February 2011. The field water cut was successfully controlled, and the oil recovery was further improved (**Figure 8**). By the end of 2017, oil recovery had increased by 5.5%.

5. Prospect of offshore chemical EOR potential

At present, the offshore chemical flooding technologies represented by polymer flooding and surfactant-polymer flooding have achieved successful field tests and applications in China, and their technical reliability and economic effectiveness have been confirmed. However, it still faces some important challenges, including how to affordably develop the movable heavy crude oil with formation viscosity above 150 mPa s, how to effectively contain the water channeling to further improve oil recovery, how to achieve efficient treatment of polymer-bearing produced fluid on offshore platforms, etc. Therefore, it is necessary to continuously explore and develop novel offshore chemical flooding technologies, such as the heavy oil activator flooding, intelligent chemical flooding, and their combination technology.

As for the field application of the offshore chemical flooding in China, the cumulative oil increment is estimated to be 8.23 million cubic meters in the three oil fields currently performed. The total economic benefit is 9.3 billion yuan, and the input-output ratio is 1:3.7. In addition, according to the present offshore chemical potential evaluation and plan, 41 reservoir units are suitable for chemical flooding in Bohai Bay, whose oil recovery could be improved by 6.53% according to prediction [27, 28]. At the same time, China offshore chemical flooding technology and field applications provide valuable and practical reference for domestic and overseas counterparts. TOTAL and Chevron have, respectively, implemented polymer flooding tests in fields in West Africa, the North Sea, and Malaysia [28]. All of these attempts will further promote the progress of offshore chemical flooding and provide technical support for enhancing oil recovery in offshore oil fields.

6. Conclusions

1. Offshore chemical flooding technology system has been preliminarily established in China including high-efficiency development model, chemical flooding agents, platform injection system, and produced liquid treatment technology.

2. Field tests and applications were successfully carried out in S, L, and J oil fields since 2003, and expected oil increment benefits have been achieved, which proved that offshore chemical EOR technology is feasible.

3. It has explored a new road for increasing the recovery rate of offshore oil field and provided a solid technical guarantee for the efficient development of offshore oil fields. With the expansion of the scale of offshore chemical flooding application at home and abroad, more and more considerable oil recovery will be achieved.

Author details

Jian Zhang[1,2], Fengjiu Zhang[1], Xiaodong Kang[1,2] and Baozhen Li[1,2*]

1 State Key Lab of Offshore Oil Exploitation, Beijing, China

2 CNOOC Research Institute Co., Ltd., Beijing, China

*Address all correspondence to: libzh2@cnooc.com.cn

References

[1] Lake LW. Enhanced Oil Recovery. Englewood Cliffs, New Jersey: Prentice Hall; 1989

[2] Guo H, Dong J, Wang Z, et al. 2018 EOR Survey in China-Part 1. In: SPE Improved Oil Recovery Conference; 14-18 April; Tulsa, Oklahoma, USA; 2018

[3] Sheng J, Leonhardt B, Azri N. Status of polymer-flooding technology. Journal of Canadian Petroleum Technology. 2015;**54**(02):116-126

[4] Sheng J. A comprehensive review of alkaline-surfactant-polymer (ASP) flooding. Asia-Pacific Journal of Chemical Engineering. 2014;**9**(4):471-489

[5] Baginski V, Larsen D, Waldman, et al. Logistic considerations for safe execution of offshore chemical EOR projects. In: SPE Enhanced Oil Recovery Conference; 11-13 August; Kuala Lumpur, Malaysia; 2015

[6] Zhou S, Han M, Xiang W, et al. Research and application of polymer flooding technology in Bohai bay. China Offshore Oil & Gas. 2006;**18**(1):1-10

[7] Zhang F, Jiang W, Shun F. Key technology research and field test of offshore viscous polymer flooding. Engineering and Science. 2011;**13**(5): 28-33

[8] Han M, Xiang W. Application of EOR technology by means of polymer flooding in Bohai oil fields. In: SPE International Oil & Gas Conference and Exhibition in China; 5-7 December; Beijing, China; 2006

[9] Zhou W, Zhang J, Feng G, et al. Key Technologies of Polymer Flooding in Offshore Oilfield of Bohai bay. In: SPE Asia Pacific Oil and Gas Conference and Exhibition; 20-22 October; Perth, Australia; 2008

[10] Zhu Y, Zhang J, Jing B, et al. Ageing effect on steady rheological behavior in oilfield water of water-soluble, hydrophobically associating polymer. Advanced Materials Research. 2013;**781-784**:431-435

[11] Xie K, Lu X, Li Q, et al. Analysis of reservoir applicability of hydrophobically associating polymer. SPE Journal. 2016;**21**(01):1-9

[12] Zhang J, Wang S, Zhu Y, et al. Study and pilot test of activator flooding for heavy oil. 26-28 March. SPE EOR Conference at Oil and Gas West Asia; Muscat, Oman; 2018

[13] Zhang J, Hua Z, Zhu Y, et al. Properties of activator with high molecular weight and its effect on viscosity reduction and desorption of heavy oil. China Offshore Oil and Gas. 2018;**30**(04):97-105

[14] Zhu Y, Zhang J, Zhao W, et al. Research on efficient polymer dissolving technology for hydrophobically associating polymer flooding on offshore platform. Applied Mechanics and Materials. 2014;**670-671**:290-294

[15] Chen W, Zhao W, Zhang J. Research on the efficient preparation method of the polymer solution in offshore oilfields. Oil-Gasfield Surface Engineering. 2017;**36**(5):20-23

[16] Zhang L, Zhang Y, Gong B. The development of a new reservoir simulator to model polymer flooding and advanced wells for enhanced heavy oil recovery in Bohai bay. In: Canadian Unconventional Resources and International Petroleum Conference; 19-21 October; Calgary, Alberta, Canada; 2010

[17] Chen H, Tang H, Duan M, et al. Oil-water separation property of polymer-contained wastewater from

polymer-flooding oilfields in Bohai bay, China. Environmental Technology Letters. 2015;**36**(11):1373-1380

[18] Xu W, Dai Z, Wen Z, et al. Problems in and measurements of the treatment of output liquid in typical offshore polymer injection oil fields. Industrial Water Treatment. 2016;**36**(7):93-96

[19] Kang X, Zhang J, Sun F, et al. A review of polymer EOR on offshore heavy oil field in Bohai Bay, China. In: SPE enhanced oil recovery conference; 19-21 July; Kuala Lumpur, Malaysia; 2011

[20] Luo X, Zhao C. Practices and experiences of seven-year polymer flooding history matching in China offshore oil field: A case study. In: SPE Reservoir Characterisation and Simulation Conference and Exhibition; 9-11 October; Abu Dhabi, UAE; 2011

[21] Kang X, Guo F, Li Y. Experience from polymer-EOR on offshore heavy oil field in Bohai Bay, China. Journal of Petroleum Engineering & Technology. 2013;**3**(1):14-19

[22] Li B, Kang X, Zhang J, et al. A systematic study of the effect of plugging on polymer flooding in W offshore oilfield of Bohai Bay. In: SPE Improved Oil Recovery Conference; 14-18 April; Tulsa, Oklahoma, USA; 2018

[23] Kang X, Zhang J. Surfactant polymer (SP) flooding pilot test on offshore heavy oil field in Bohai Bay, China. In: SPE Enhanced Oil Recovery Conference; Kuala Lumpur, Malaysia; 2-4 July; SPE-165224-MS; 2013

[24] Lu Q, Ning Y, Wang J, et al. Full field offshore surfactant-polymer flooding in Bohai Bay China. In: SPE Asia Pacific Enhanced Oil Recovery Conference; 11-13 August. Kuala Lumpur, Malaysia; 2015

[25] Ma K, Li Y, Wang L, et al. Practice of the early stage polymer flooding on LD offshore oilfield in Bohai Bay of China. In: SPE Asia Pacific Enhanced Oil Recovery Conference; 11-13 August; Kuala Lumpur, Malaysia; 2015

[26] Xie X, Kang X, Zhang X, et al. Chemical flooding potential evaluation method in offshore heavy oil fields and its application. China Offshore Oil and Gas. 2016;**28**(1):69-74

[27] Yu B, Xia Z, Du Q. Evaluation method and software development for oilfield chemical flooding potential. Journal of Yangtze University (Natural Science Edition). 2016;**13**(26):46-49

[28] Morel DC, Jouenne S, Vert M, et al. Polymer flooding in deep offshore field: The Dalia Angola case. In: SPE Annual Technical Conference; 21-24 September; Denver, Colorado, USA; 2008

Chapter 7

CO_2-EOR/Sequestration: Current Trends and Future Horizons

Erfan Mohammadian, Badrul Mohamed Jan, Amin Azdarpour, Hossein Hamidi, Nur Hidayati Binti Othman, Aqilah Dollah, Siti Nurliyana Binti Che Mohamed Hussein and Rozana Azrina Binti Sazali

Abstract

The use of carbon dioxide (CO_2) as an improved oil recovery (IOR) method has been a common practice in petroleum engineering. In this chapter, various technical aspects of application of CO_2 to increase oil recovery are discussed. From the required laboratory tests prior to field applications to postinjection monitoring of injected plume, the required onshore and offshore facilities, the environmental considerations, and challenges concerning the application of CO_2 for EOR purposes have been covered in this chapter. Moreover, the emerging methods and industry trends in applications of CO_2 for EOR will be discussed. The second part of this chapter is dedicated to CO_2 sequestration as a method to mitigate CO_2 emitted due to the anthropogenic activities. CO_2 sequestration is the injection of large quantities of CO_2 into underground reservoirs (oil and gas, aquifers, and coal deposits) where it can be securely and permanently stored.

Keywords: improved oil recovery, enhanced oil recovery, CO_2 sequestration, CO_2-EOR, CO_2 miscibility

1. Introduction

Current dilemma faced by the United States in lowering the dependency on foreign energy source and curbing emissions of greenhouse gases has brought light to carbon dioxide (CO_2)-enhanced oil recovery (EOR) method [1]. Oil recoveries using natural (primary) mechanisms rarely exceed 20% of the original oil-in-place (OOIP) [2]. Secondary methods of recovery often add few percentages to the above figure [2]. Hence, companies would welcome any methods that could lead to an increase in the production of the postwater/gas injection trapped oil in the reservoir. Injecting CO_2 into oil reservoirs to improve the recovery of oil on commercial (field) scales has been practiced since nearly half a century ago. The idea of CO_2 flooding first emerged in 1930s, and more laboratory and field studies were conducted between 1950s and 1970s. Most of these implementations have been in North America. However, some small-to-large scale CO_2 injection projects have been reported in other parts of the world [3, 4].

In this method, CO_2 is usually injected as a supercritical fluid. When injected at or above the critical point of pressure and temperature, supercritical CO_2 can maintain the properties of a gas while having the density of a liquid. In this state, oil could be more efficiently mobilized from the depleted reservoir due to the improved volumetric efficiency. Conventionally, CO_2 injection method is usually applied to the reservoirs with oil gravity less than 25 [5]. As CO_2 is injected into the reservoir, the miscible CO_2 will blend thoroughly with the oil in a manner that the interfacial tension between these two fluids becomes zero. The other mechanisms of CO_2 by which the oil recovery is improved are the dissolution of CO_2 in oil, swelling of oil, and eventually reduction of the viscosity of oil.

Since 2002, as a consequence of Kyoto protocol and imposing of the carbon tax, CO_2 sequestration as a method to mitigate the high concentration of CO_2 in the atmosphere has received a lot of attention [6, 7]. However, the lack of economic incentives has been the biggest hindrance to industrial field-scale application of CO_2 sequestration. Emerging in the last decade, CO_2-EOR was proposed as a method to add economic benefits of CO_2 injection to mature oil fields to the environmental merits of CO_2 sequestration [6]. Therefore, considering the large amounts of research dedicated to CO_2-EOR, it is expected that in the near future, more field applications of this technology will be implemented globally. In this chapter, the phase behavior and hydrocarbon miscibility of CO_2 is discussed in detail.

2. Hydrocarbon miscibility

With decline in overall production levels from mature oil fields, oil companies have turned to enhanced oil recovery (EOR) techniques as a way of maximizing output. The more commonly applied technique is gas injection or miscible flooding. Miscible flooding is a commonly used term used to describe gas injection processes. This involves the displacement of oil that aids in maintaining original reservoir pressures by reducing the interfacial tension that exists between the oil and gas phases. This acts by removing the interphase between the two fluid phases, and commonly used gases include CO_2, natural gas, and nitrogen, with CO_2 being the most prominently used gas. Research on the use of CO_2 has been ongoing since the 1950s, continuing into the 1960s [8–10]. And the advantage that CO_2 injection brought about was noticed as the increase in reservoir pressure that resulted in higher oil production due to the driving force provided by it. In its infancy, research on CO_2 showed it to be immiscible with oil at reservoir pressures, but it was later discovered that under certain conditions of temperature, pressure, and oil composition, the carbon dioxide becomes enriched and becomes miscible with oil [11]. The pressure required for CO_2 gas to attain miscibility in oil is also much lower than methane gas. The term "miscible flooding" has been adopted as the conventional phrase used to describe the process of gas injection.

The main advantages of using CO_2 in this process include the following:

- The miscibility of CO_2 with oil as highlighted earlier.
- It is a cheaper source of gas than other alternatives.
- By injecting it back, CO_2 capture is also achieved.

From review of various literatures, an indication of the suitability of CO_2 as an excellent solvent for EOR in onshore fields of Canada and the USA can be deduced. And from experience garnered worldwide by operators, CO_2 flooding increases oil extraction by between 7 and 15% of oil initially in place. And also, it reduces the amount of CO_2 in the atmosphere and greenhouse gases in general [12].

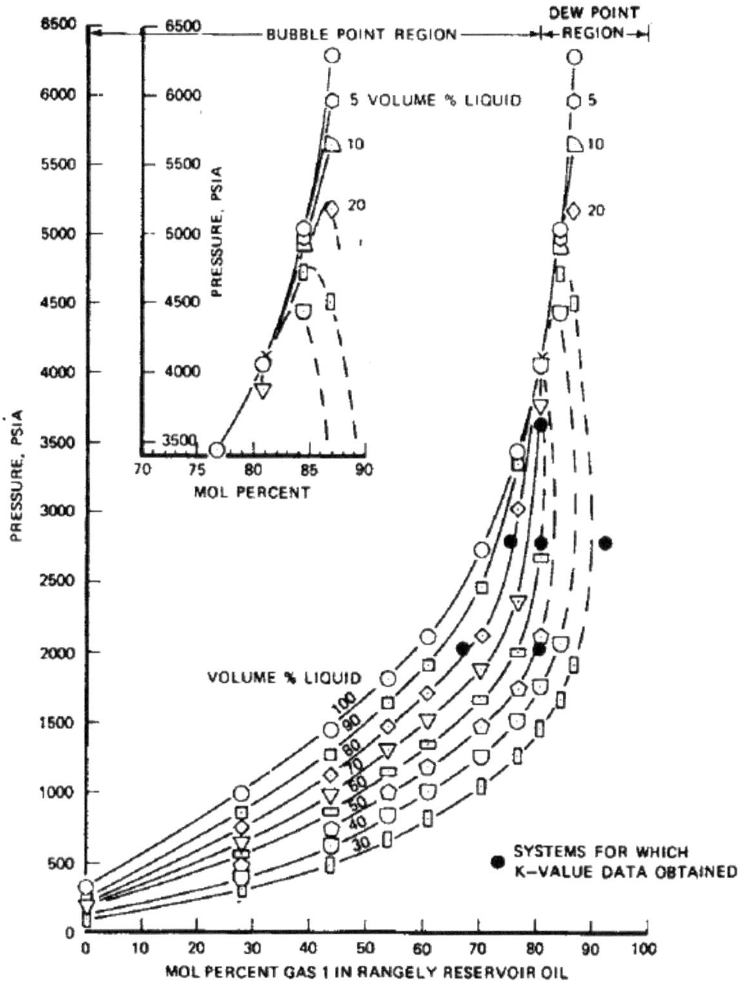

Figure 1.
Pressure composition diagram—gas 1 system for Rangely oil: 95% CO_2 and 5% CH_4 gas at 160°F [13].

Challenges associated with CO_2-driven EOR include technological, economical, and supply. For example, a long pipe network is usually required to transmit CO_2 from source to the field. High-pressure compressors are also another essential requirement in the injection process. Therefore, all these factors have to be assessed and weighed in relation to the extra oil recovered to determine if it is profitable.

In recent times, newer techniques such as water alternating CO_2 injection and simultaneous water and CO_2 injection have been developed, and they are determined to increase efficiency of oil recovery at lower costs.

A recent study commissioned by the Congressional Research Service shows that, theoretically, carbon-capture technology could remove as much as 80–90% of CO_2 from emissions.

The minimum miscibility pressure (MMP) of CO_2 and NGL for conventional reservoirs are few hundreds psia higher compared to unconventional reservoir due to difference in pore size according to a study by Teklu et al. [13].

2.1 MMP in nanopores, fluid properties, and phase behavior

The deviation between the nanopore phase behavior from bulk (PVT cell) properties was studied [14–16]. The bubble point and dew point pressure, interfacial tension (IFT), and minimum miscibility pressure (MMP) between injection and reservoir fluid change in nanopores due to small pore confinement effect [13]. MMP was calculated by including capillary pressure and critical property shifts in confined pores using multiple mixing cell (MMC) algorithm of Ahmadi and Johns [17].

Phase behavior is important in the design of a variety of EOR processes, for example, surfactant/polymer processes and gas injection processes. The process of reducing interfaces between oil and displacing phase and hence removing effect of capillary forces between injected fluid and the oil is called miscible displacement. During the gas injection process, the required miscible-displacing fluid is generated by mixing the injected fluid with oil in the reservoir. Phase behavior of gas/oil systems is summarized in the pressure-composition (p-x) diagram. A work by Graue and Zana [18] summarizes the result for CO_2 injection in the Rangely field, Colorado. The physical property date was obtained from constant composition expansion (CCE) to determine the phase envelope (bubble point and dew point envelope) and vapor/liquid equilibrium experiment (VLE) to yield vapor/liquid equilibrium constant (K-values). The phase behavior of Rangely reservoir oil with different gases' composition at reservoir temperature of 160°F showed that critical and saturation pressures of the injected gas/reservoir oil system were increased substantially by 10 mol% N_2 in the injected gas. The phase behavior data showed solid phase precipitation that amount for 2–5% of the reservoir oil [18]. **Figure 1** illustrates the pressure composition diagram of Rangely oil containing considerable amounts of CO_2.

3. CO_2 miscible injection method

The oil displacement process is classified into two types depending on the method on which miscibility is achieved. These are classified as first-contact miscible (FCM) and multiple-contact miscible (MCM). In the FCM process, a small quantity of a primary slug that is miscible with the oil is initially injected; afterwards, a larger quantity of a less expensive slug is injected. The size of the slugs injected is determined by the costs. Under ideal conditions, the two injected slugs should be miscible; thus, at both the leading and trailing edges of the primary slug, the phase behavior has to be monitored. In the case of these slugs being immiscible, a residual saturation of the primary material will be trapped in the displacement process. While during the MCM displacement process, miscibility in the reservoir is generated through in-situ composition changes due to multiple-contacts and mass transfer between the injecting fluid and oil present. These MCM processes are classified as displacements using vaporizing gas (lean gas), condensing and condensing/vaporizing gas (enriched gas), and CO_2.

3.1 Vaporizing gas drive mechanism

A relatively lean gas is gas containing a little low molecular weight hydrocarbon (or inert gases like nitrogen) and methane making up the rest of the composition. The schematic of the CO_2 (**Figure 2**) miscible process shows the transition zone between the injection and production well [19].

After injection, its composition gets changed as it moves through the reservoir in the process becoming miscible with the original reservoir oil. This means that through multiple-contact the composition of the injected fluid is enriched, and intermediate components are vaporized into the injected gas. And at some point under the appropriate conditions, the enrichment reaches a level where the injected gas becomes miscible with oil in the reservoir. It is from this stage of the process, under ideal conditions, that displacement is said to occur [20–22].

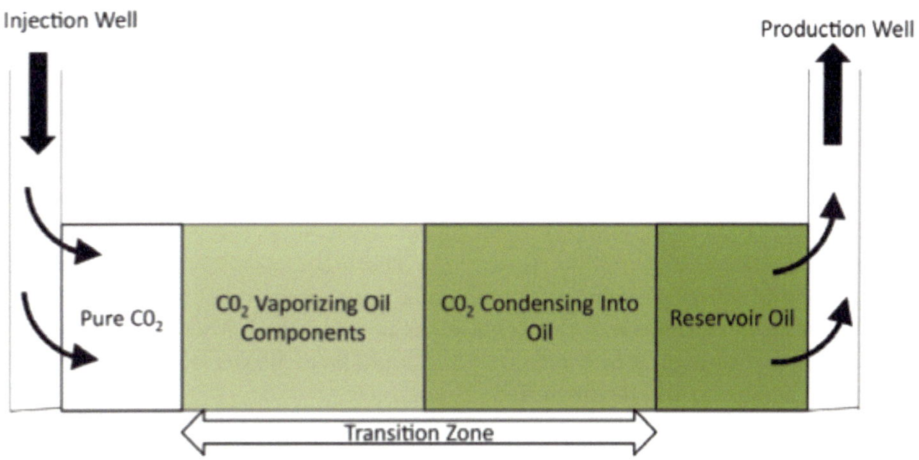

Figure 2.
Mechanisms of CO_2 injection for EOR [19].

When using a condensing or enriched gas as the injecting fluid, the process is more expensive because the fluid tends to contain a high concentration of intermediate-molecular-weight hydrocarbons. This process entails enrichment of the reservoir oil that first comes in contact with the injection fluid. Thereafter, hydrocarbon components from the fluid are condensed into the oil, giving it the name condensing process. Under ideal conditions, this oil is sufficiently changed in composition such that it becomes miscible with increased fluid injection and miscible displacement thus occurs. This process can be operated at a lower pressure than the vaporizing process [23–26].

It has been a long held notion that the enriched-gas process is operated mechanically, as highlighted in the previous paragraph. However, it has now been discovered that it is more often a combination of condensing and vaporizing mechanisms. The lighter components of the injected gas (C2 through C4) tend to condense into the reservoir oil as previously highlighted. While the middle intermediate components (C4+) become vaporized from the oil and absorbed into the gas phase, this prevents the development of miscibility between fresh injected gas and enriched oil at the entry point of the injection process (the oil becomes heavier). Further into the injection process, the light intermediates in the gas condensate into the oil, and this leads to the oil becoming saturated. As for the middle intermediate, vaporization continues due to the slight enrichment of the injected gas. When the condensation/vaporization process proceeds further downstream, the gas becomes more enriched due to contact with the oil. And the enrichment is said to occur at the point where the gas "nearly" becomes miscible with the original reservoir oil, ensuring a more efficient displacement process, even though miscibility is never fully developed (i.e., the two phases are never fully miscible in all proportions) [20, 27–29].

CO_2 is not miscible with most crude oils at first contact under normal reservoir conditions. However, at some ideal conditions of temperature, pressure, and composition, miscibility is expected to occur through multiple contacts. Overall, the process behavior is analogous to the vaporizing process. Under some conditions, the phase behavior can be more complex, having two liquid phases, or two liquid phases in addition to a vapor phase.

3.2 CO_2-EOR injection consideration

There are two main groups of considerations for CO_2-EOR, namely technical and economical (fiscal) considerations. CO_2-EOR injection technical consideration involves a complex engineering and differs from reservoir to reservoir. A detailed description of reservoir field and prospect of miscibility must be taken into account before considering CO_2-EOR injection. Usually, the key parameters used in the technical consideration are remaining oil in place, minimum miscibility pressure, reservoir depth, oil API gravity, and formation dip angle.

In offshore fields, there are more factors that need to be considered. Firstly, the separation of CO_2 from the produced gas is the ideal choice if the CO_2 source is not in the vicinity of the field. Next, at high CO_2 concentrations, there is a need for the facilities and operation to process the gas [21]. This is because CO_2 becomes acidic as it is injected into the (formation) water and causes corrosion of the equipment in the offshore environment. If the fields are using CO_2-WAG processes, then the facilities need to be compatible with acid that could be generated so that the corrosion in the facilities could be prevented [20–22].

Verma et al. [29] studied the parameters that affect the efficiency for increasing the production of methane gas on Marcellus shale and concluded that the gas production can be increased by 7% with the optimal spacing between injection and production well. It can be concluded that the natural fracture permeability is the dominant factor to improve the production of methane. As the fracture half-length increases, the methane production increases and the possibility of CO_2 breakthrough also increases. The down side of this process is the cost as well as a high risk of leakage and the field pollution. This is due to the fact that the injecting of CO_2 can degrade the gas production as a result of the mixing initial gas in place with the injected CO_2 [30]. Due to the miscibility of CO_2 and the natural gas, their physical properties were potentially ideal for reservoir re-pressurization. For instance, CO_2 has higher density and lower mobility ratio compared to methane. Hence, CO_2 will sink in the reservoir; this can stabilize the displacement process between the injected CO_2 and the methane initially in place.

Reservoir heterogeneity and solubility of CO_2 in formation brine could also play a major role in causing early CO_2 breakthrough to the production wells. The latter could be delayed, by re-pressurizing the reservoir [26, 31]. Generally, due to the benefits of CO_2 injection to gas reservoir, CO_2-EGR could be potentially efficient and therefore an attractive option in spite of a bigger investment required as compared to CO_2 injection into oil reservoirs. Nevertheless, it can extract more hydrocarbons as compared to oil reservoirs.

The reservoirs must be subsequently screened for economic consideration based on standardized capital costs and operation expenses that are representative of the reservoirs under consideration. Wei et al. [27] found that the total crude oil recovery potential along with CO_2 storage resource and net income for enterprises can be increased if the price of crude oil is high and the price of CO_2 and tax is low. The cumulative cost-effective oil production varied between 0.3 and 1.3 billion tons (2.1 and 9.1 billion barrels). This is consistent with research reported from Appalachian basin region, which suggests that CO_2-EOR may be economically feasible in the study area when oil prices are $70/STB or higher [28, 32]. However, the economics of onshore CO_2-EOR will face an undesirable impact due to complex geological properties, high viscosity of crude oil, high royalty rates, technology limitations, and the lack of incentives for CO_2-EOR projects. Overall, a miscible CO_2-EOR process is preferred considering all the technical and economical evaluations as detailed as possible.

4. The CO_2 injection and sequestration facilities

In this chapter, the facilities that are required for CO_2 injection and CO_2 sequestration application both in the offshore and onshore environment are discussed.

4.1 Onshore surface facilities

For CO_2-EOR, the facilities required are almost the same with those that are required in water-flooding process that includes gas phase (CO_2 and natural gases) gathering lines, CO_2 metering, and distribution lines that are also required in the designing of the facilities for the CO_2-EOR operations [33, 44]. However, there are three basic elements that differentiate the two processes. They are as follows:

1. Extraction: CO_2 gas is extracted from the increasingly rich CO_2 separator gas due to its breakthrough in producing wells.

2. Processing: It is the process of purifying of CO_2 down to specification upon extraction from the separator gas and CO_2 undergoing dehydration prior to compression.

3. Compression: CO_2 undergoes compression to raise its pressure for injection.

One of the preliminary considerations for CO_2 EOR facilities is the incorporation of the flue gas CO_2 recovery plants, CO_2 compression/dehydration unit, CO_2 pipelines, CO_2 injection wells, and a facility for separating CO_2 from associated gas as can be seen in **Figure 3**. Macon extensively discussed the details of Levelland Texas, which is one of the earliest projects of CO_2 injection, from aspects of the design and operation of the facilities [33, 34].

4.1.1 CO_2 transport, storage, and refrigeration

CO_2 transport is the first operational aspect to take into consideration in a CO_2 injection project. CO_2 transportation by trucks or rail cars may be deemed inefficient and costly if the supply source is far away from the injection site [34]. Moreover, since CO_2-EOR requires millions and billions of tons of CO_2, such methods of transportation are simply impractical. In the later projects, the CO_2 is transported via pipeline system as a dry gas, separated from any existing water injection system.

The CO_2 will be stored in liquid form at approximately −18°C and ~300 psi and this can be achieved by utilizing a refrigeration unit that consists of skid mounted compressors and refrigerant condensers. The refrigeration system when added with heavy insulation will minimize the cost of the product normally lost through relief valves as the liquid warms during prolonged storage. The CO_2 gas is dehydrated as its dew point is lowered to a temperature lower than 0°C and compressed up to its critical pressure before being fed to the reservoirs via a CO_2 pipeline [35, 36].

4.1.2 CO_2 distribution and injection

In most cases, CO_2 is distributed throughout the system and facilities as a dry gas through a trunkline system [33, 35, 37] using bare carbon steel systems as there is no concern for corrosion when the pumping pressure is set at around 1800–2400 psi [37]. However, precautions should be taken by moving the flange valve and tying the valve above ground to avoid leaks and the dry ice formation if loose. Blowdowns consisting of a buried flanged blowdown valve, a blowdown stack, and a line blind are installed (refer **Figures 3** and **4**) to remove excess CO_2 impurities in the pipe.

Special consideration should be taken into account for Christmas tree if water alternating gas (WAG) method is applied. If water and CO_2 are alternately injected, the part of the system will be vulnerable to corrosion as high-pressured CO_2 will come into contact with water and form an acidic solution [36].

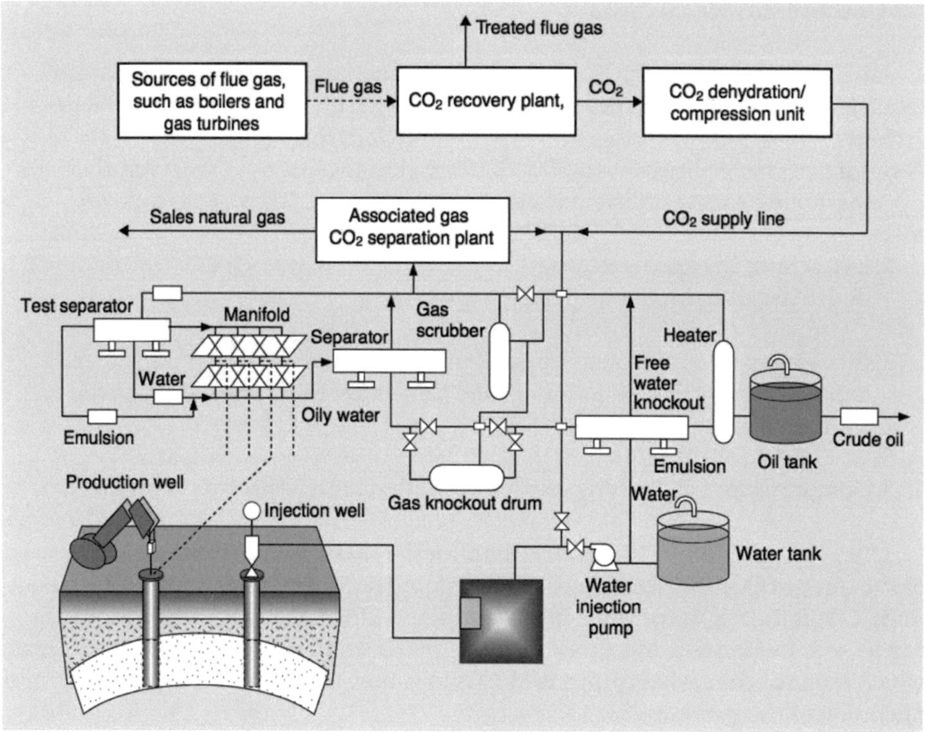

Figure 3.
A typical system utilized in CO_2-EOR [4].

At the CO_2 well head, an additional tee is necessary to be installed, which allows the high-pressure CO_2 stream to be closed off away from the well head, which in turn will increase the safety of workovers and similar operations (refer **Figures 5** and **6**). No modifications are required in terms of artificial lift equipment or the well head equipment during the implementation of the CO_2 project mainly due to economic reasons and uncertainties of design parameters [37].

Injected CO_2 eventually recycles back. The recycled stream may contain H_2S [43]. While this may raise a cause for a higher possibility for stress cracking, unlike most systems, the recycle stream will be sufficiently dehydrated, so the need for protection can be minimized through mill analysis and inspection.

4.1.3 Field production facilities

Modifications to the well head were implemented in the ADCO's Wasson ODC Unit, which include 80 nipples, 2000 psi-rated ball valves on the tubing head, and new elastomers in both the secondary seal and tubing slip seal. For beam pumped wells, the blowout preventer elastomers are also changed. For electrical submersible pumps, the tubing valve is being replaced with a 2000 psi-rated gate valve. Many equipment changes and tests are being conducted with respect to all artificial lift equipment. Pre-CO_2 injection flowlines are to be replaced with higher pressure rated fiberglass pipes to increase line capacity, and not for corrosion resistance [38].

4.1.4 Remote production headers

The remote headers and satellite battery headers are responsible to collect, or centralize, production from numerous wells and to provide individual well test

DOI: http://dx.doi.org/10.5772/intechopen.89540

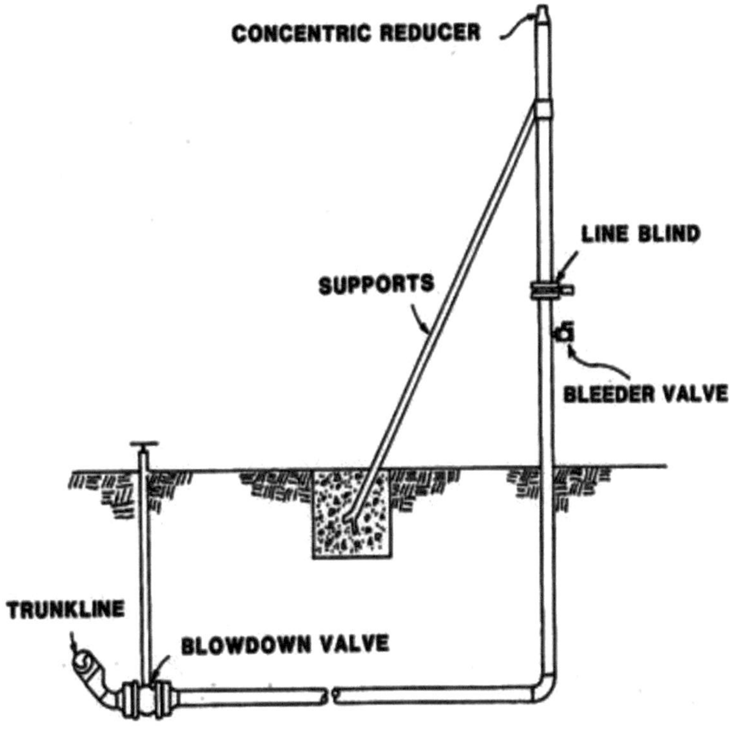

Figure 4.
CO_2 trunkline blowdown station [5].

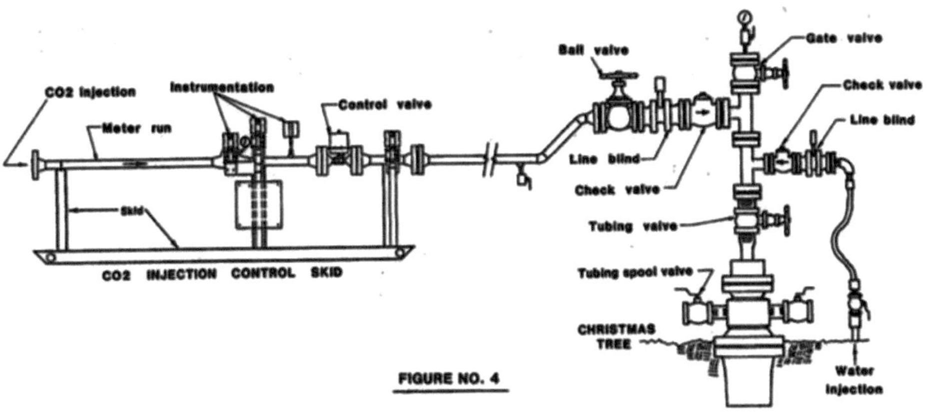

Figure 5.
CO_2 injection well head [5].

facilities for oil, water, hydrocarbon gas, and CO_2 gas production. Remote headers are operated at pressure range of 30–35 psi, while the satellite batteries' headers operate from 25 to 30 psi, whereas satellite batteries are to remove the majority of the hydrocarbon and CO_2 gas from the produced fluids and to pump liquids to the fluid gathering, and obtain well tests [39]. Each satellite battery includes a header, test separator, production separator, and transfer pumps.

The specific design criteria for satellite batteries include the following:

1. Minimum of two phase production separators per satellite

2. A pump from separator system

3. Three phase test separators per ten wells

4.1.5 Separation process

Surface facilities in CO_2-EOR requires recovery of CO_2 and reinjecting it back to the well, with which CO_2 release to the atmosphere can be minimized as well as purchasing cost of additional CO_2 can be reduced. Typical surface facilities for CO_2-EOR are gas separation. Water treatment of CO_2 compression and injection and also dehydration can be seen in **Figure 7**.

Dehydration column removes the moisture content of the gas stream by using the contact with lean glycol in the upper part of the 14 tray column and the gas must be cooled first by the air cooler. Rich glycol is extracted from the bottom of the

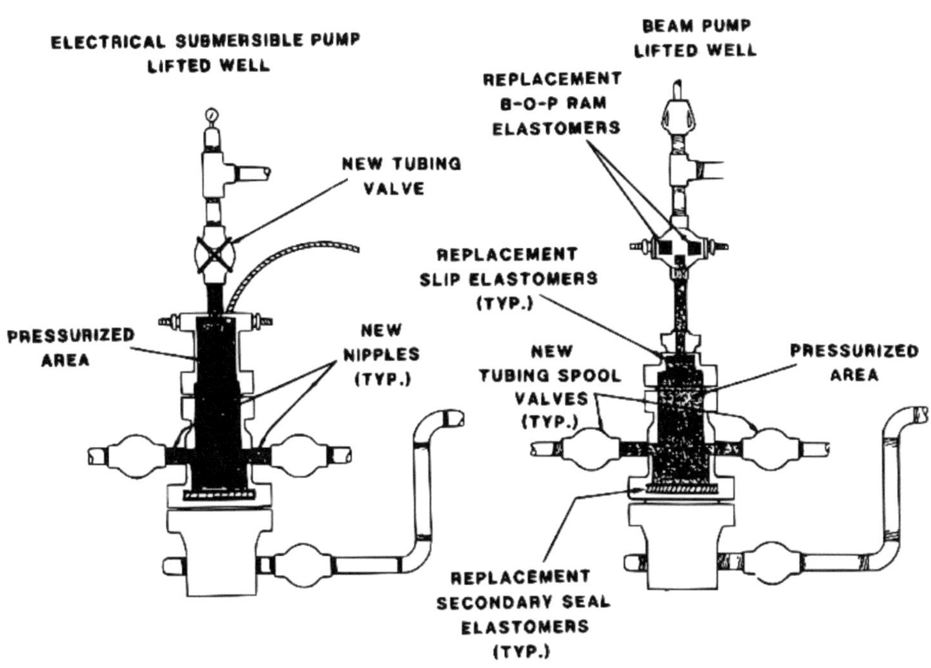

Figure 6.
Modifications on CO_2 production well head [5].

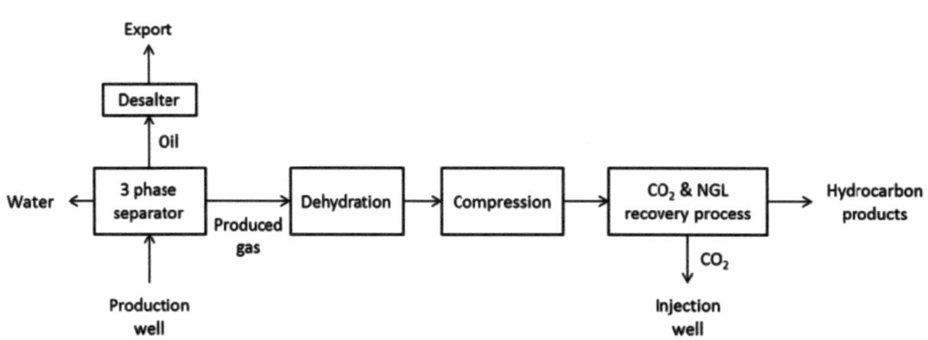

Figure 7.
Typical surface facilities for CO_2-EOR [21].

column and directed to the regeneration system, and dry gas passes through from the top to the compressor package.

A slug catcher is to catch any water as water will be continuously dropping, which may unload in slugs. Then, three-stage reciprocating compressor is used to increase the pressure of CO_2 from 700 kPa(g) to approximately 10,000 kPa(g). At the first stage reciprocating compressor at 3000 kPa the CO_2 is dehydrated. Afterwards, CO_2 liquid is trucked to the site to supplement injection requirement and stored in refrigerated bullet. Then, the cryogenic triplex pumps are used to pump the CO_2 liquid to injection pressure. After that, the liquid and compressed gaseous CO_2 are commingled in a mixer [40]. A minimum well head temperature must be maintained so as not to freeze the inhibited water in the well annulus. Should the injection compressor be shut in, liquid CO_2 is warmed by a trim heater.

4.2 Offshore CO_2-EOR facilities

The majority of CO_2-EOR projects are all similar in terms of facilities to those in the offshore. The following sections discuss the various equipment and facilities required at different phases of a typical offshore CO_2 injection project.

4.2.1 Pig launcher and steel pipeline

Depending on the CO_2 stream composition, weight, temperature, and pressure, the outline of pig launcher ought to be in certain standards and suitable measures (e.g., DNV OSF-101). Pigging is usually used for dispatching new or re-appointing existing pipelines. A reconditioned or new pipeline needs to be hydrotested to guarantee pressure integrity. Frequent inspection for any degradation and verification of the effects of dry CO_2 stream on the pipeline must be done. In any case, since the CO_2 will be exceptionally dry, it is likely that future pigging operations will be less successive due to subsea situation. Normally, the pig traps are not fitted; thus, portable pig traps can be considered in the design [40].

Piping diameters and thickness are the main concern in designing the facilities in which for the purpose of CO_2-EOR the pipeline could have two diverse design requirements. If the storage complex is already pressurized over the required level to keep the CO_2 as fluid in injection wells, then the project may pick to utilize high-pressure liquid pipe. In other cases, for instance, the Hewitt Field model [11], the pressure inside the field would bit by bit be expanded using gaseous CO_2 in the transport system until the field pressure can maintain liquid CO_2 in the injection wells. During that time, the transport system would be changed over to a higher pressure fluid pipeline.

The pipeline system in onshore system is typically below 120 atm (1740.45 psi) and logically the pressure in the offshore CO_2 pipeline is higher. In offshore frameworks, the CO_2 is being pumped under long distances; thus, there is much pressure loss due to friction. With trunkline pressure between 150 and 250 atm, the CO_2 may be injected into geological formation without further boosting of pressure. The pressure will be affected by the distance of onshore facilities and offshore storage complexes, which may affect the wall thickness and pipeline diameters [21, 41].

Due to the danger of transporting high-pressure liquid CO_2, the transportation will be done in liquid and subsupercritical. To diminish the risk in onshore pipelines, it is likely that these will be at pressure lower than the required for offshore transport and ought to be expanded at a coastal booster station before going to offshore [42–44]. Besides, offshore pressure booster may be required if there is excessive pressure drop between onshore booster station and offshore storage complexes or if the pressure is inadequate to inject directly into reservoir [45].

Injection rate	Pressure	Temperature	CO_2 phase
Range: 10–48 t/day	Well head 1015.26 to 1595.42 psi	Well head 32°C	Kept to be "supercritical phase" (at well bottom)
Ordinary rate: 20 to 40 t/day	Well bottom max. 2755.72 psi	Well bottom 48°C	

Table 1.
Operation conditions of injection facilities in Nagaoka, Japan [42].

To avoid two-phase flow that leads to critical damage and cavitation, the pressure downstream of an offshore booster pump should be over bubble point [46].

4.2.2 CO_2 injection

CO_2 might be injected directly into the reservoir if the pipeline pressure is adequate. However, it is likely to bring CO_2 onto platform for control pressure or to lift its pressure; certain design considerations should be considered. Below are the injection facilities' operation conditions and the process flow diagram of field test of CO_2 injection in Nagaoka, Japan [43]. Some operation details of injection facilities in Nagaoka are summarized in **Table 1**.

4.2.3 Risers, emergency shut down valve (ESDV), compressors, and pumps

Risers are the piping that transports the fluid between the offshore platform and the seabed. Flexible risers are used especially on floating production installations. ESDV is placed between the moving pipe infrastructure and the riser to the platform as a safeguard gadget to guarantee no leakage of CO_2 when there is failure in platform. It is likely found on the seabed where there is the possibility of heavy things to be dropped on the pipeline underneath during the lifting operation work. Moreover, it is designed to counter any structural failure on the platform and any upstream failure where it will cause the valve to close.

Pressure issue related to CCS projects may require extra pumping units due to higher pressure required over long distance. In the event that the pressure drops along offshore pipeline, usually pumps would be placed on an offshore platform [44].

4.2.4 Fluid separation

Due to the nature of CO_2-EOR patterns, the water production is high, thus leading to the need for large separation capacity with inlet separators dominated with water rather than oil. Moreover, separation is harder because of scale, emulsion, ESP or gas lifting, and asphaltenes. A large CO_2 reinjection compressors are needed due to high CO_2 production that resulted from back produced CO_2 in the system.

5. CO_2 sequestration

Since the industrial revolution, concentration of CO_2 and other greenhouse gases increased due to burning of fossil fuels. The measured atmospheric concentrations of CO_2 are 100 ppm higher than preindustrial levels [46]. According to report published by the Global Monitoring Division (formerly CMDL) of the National Oceanic and Atmospheric Administration, concentration of CO_2 in the atmosphere has increased by 26% from less than 320 ppm in 1960 to 405 ppm in 2017 [47]. Fossil fuels provide 86% of the world's energy and this amount is responsible for around 91% of CO_2 emissions. There is almost a global agreement about taking responsible actions in implementing energy sources in the years to come [48].

Subsurface CO_2 sequestration is one of the various options for reducing the CO_2 concentration in the atmosphere resulting from anthropogenic activities. CO_2 sequestration potentially provides around 25% of the required mitigation to global emissions, which can delay global warming to an acceptable extent [48]. Subsurface sequestration captures CO_2 at the point of its emission and injects large amounts of it deep into subsurface formations where it can be stored permanently [49].

5.1 CO_2 sequestration repositories

Several types of subsurface repositories may be utilized for sequestration of CO_2. CO_2 could be safely sequestrated in subsurface formations such as deep saline aquifers, coal bed methane (CBM), and depleted hydrocarbon reservoirs. Due to known geological formation and existence of seal traps, CO_2 may be more safely sequestrated in depleted oil and gas reservoirs as compared to saline aquifers and coal bed methane reservoirs. On the other hand, the abundance and higher storage capacity are two major motivations for sequestration of CO_2 in saline aquifers. **Figure 8** illustrates CO_2 sequestration in various underground repositories. International Energy Agency (IEA) estimated global geological sequestration (storage) potential of 400–10,000 Gt for saline formations and 900 Gt for depleted oil/gas fields [50]. CO_2 sequestration requires comprehensive knowledge of characterization and behavior of CO_2, rock and fluid interactions, as well as operation conditions in the geological formation of interest.

5.2 Mechanisms of sequestration

There are several mechanisms involved in the sequestration processes. In a typical CO_2 sequestration, some of the injected gas dissolves in the formation water (solubility trapping), some may be trapped as residual gas saturation (nonwet trapping), and some may react with host minerals to precipitate carbonate, i.e., mineral trapping.

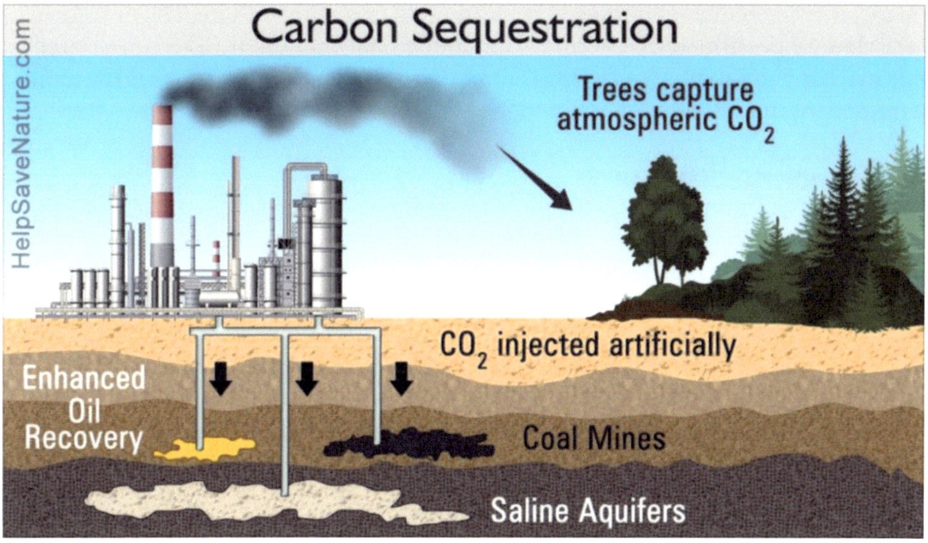

Figure 8.
CO_2 sequestration in various geological settings [from helpsavenature.com].

In trapping mechanism, the injected CO_2 is trapped in reservoirs in a manner similar to natural gas. Further vertical movement of natural gas (similar to CO_2) is hampered by cap rock, which is impermeable. Although combination of all of the sequestration mechanisms render CO_2 immobile in the geological repositories, the structural/stratigraphic and residual fluid mechanisms have the most dominant and imminent effect on trapping or retaining CO_2 in aquifers [51]. This mechanism is mainly governed by density of injected CO_2. The density difference between the injected CO_2 and brine determines further movement of CO_2 plume to rise or sink.

In nonwet trapping, once the supercritical CO_2 is injected into the formation, it relocates fluid as it passes through the porous rock. As CO_2 continues to move, some of the CO_2 is left as disconnected droplets in the interstices due to interfacial forces. This process occurs when relative permeability to nonwet phase, i.e., CO_2, becomes zero; nonwet phase therefore is rendered immobile assuming the formation is water-wet. Just like trapping of oil droplets (as nonwetting phase) in the pores containing wetting-phase (being brine), CO_2 fills the interstices between pores and is trapped as discontinuous phase. The phenomenon is largely dominated by interfacial tension between the phases and wetting characteristics of the surface [52, 53].

Dissolution of CO_2 in water is another important process responsible for sequestration of 20–60% injected CO_2 in the geological formations. Dissolution mechanism occurs during migration of CO_2 along its pathway in the injected formation. Over time, the injected CO_2 dissolves into the formation brine, increasing its density. As a result, CO_2-saturated brine sinks slowly and does not reach the surface. Moreover, the dissolution of injected high-pressure CO_2 is in the formation brine acidifies the indigenous formation water [10]. Estimating capacity of this mechanism requires reservoir simulation and knowledge of CO_2 supply ratio and injection rate, rock/fluid properties, and reactions [54].

In CO_2 mineralization, CO_2 reacts with minerals in rock to form stable components such as carbonates and aluminosilicate. It occurs along the migration pathway of CO_2 into reservoir. Both rate and magnitude of reaction are dependent on the presence of reactive minerals [52] and formation water chemistry [55, 56]. Effective time for mineralization may vary from 500 to 1000 years. However, mineralization can give rise to precipitation of certain minerals and it leads to blockage of pore throat, thereby reducing permeability leading to loss of injectivity. The process is very slow and confined CO_2 becomes immobile. The amount of CO_2 sequestrated by this mechanism can be significant. Knowledge of mineralogy of a rock is the main requirement in predicting the behavior of CO_2 in this mechanism.

5.3 CO_2 sequestration capacity

The estimation capacity can be calculated using:

$$G_{CO_2} = A \times h \times \Phi \times \rho \times E \tag{1}$$

where G is the volume of CO_2, A is the area, h is the thickness, Φ is the porosity, and E is the efficiency factor for the CO_2 sequestration operation. The abovementioned parameters are mostly in the following range: mostly within the following range in physical parameters:

- Areal extent of worldwide sedimentary basins (A): 70–80 million km^2
- Aquifer thickness (h): 50–400 m

- Porosity (Φ): 0.05–0.30
- CO_2 solubility (S): 20–80 kg/m^3; efficiency factor (Es): 0.01–0.5 (*0.0001–0.01)
- CO_2 density (ρ): 400–800 kg/m^3; efficiency factor (Ef): 0.01–0.03 (*0.0001–0.0006)
- Eq. (1) can be further modified to account for CO_2 sequestration capacity that is coming from each trapping mechanism [51].

6. CO_2 injection in unconventional reservoirs

Conventional reservoirs are oil and gas reserves that could be found in discrete accumulation of pools. Therefore, the hydrocarbon can be easily recovered through classic exploration techniques and vertical or deviated wells. Unconventional reservoirs on the other hand could be defined as a reservoir that requires out-of-the-ordinary and hence complicated techniques of recovery as compared to the conventional oil and gas reservoirs [57, 58]. The main reasons why such reservoirs are getting considerable attention are the depletion of conventional sources and huge energy demand. **Figure 9** shows the unconventional reservoirs that can be potentially produced for recovery of hydrocarbon. Tight-gas sands, gas and oil shales, coalbed methane, heavy oil, tar sands, and gas-hydrate deposits are among the most anticipated reservoirs. These reservoirs often necessitate complex recovery

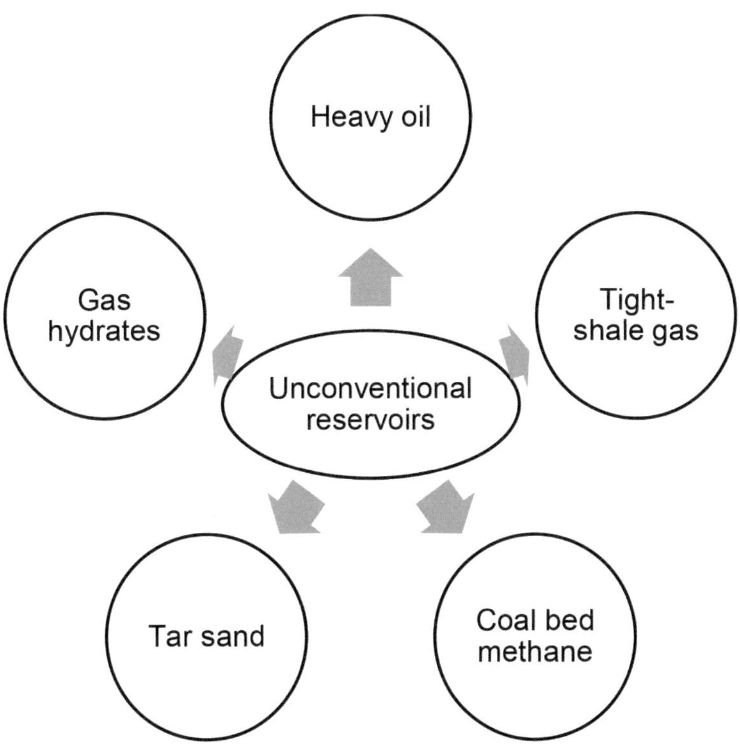

Figure 9.
Common unconventional hydrocarbon reservoirs.

solutions such as stimulation treatments or thermal recovery methods and particular process facilities. Moreover, those requirements should be technically and more importantly economically viable [59].

6.1 Shale reservoirs

Shale gas reservoir is referring to unconventional reservoirs that produce natural gas. Shale gas reservoir has received a lot of attention due to the potential reservoir in supplying clean burning energy and the way it copes with the depletion of conventional reservoirs [58]. However, at a certain time, the production of shale gas well decreases rapidly; thus, an enhanced gas recovery method has been aiming to improve the recovery from shale gas reservoirs.

In shale reservoir, methane (CH_4) is adsorbed initially onto the surfaces of matrix particles and natural fracture faces and is stored in the matrix limiting its effective extraction [59]. Although large amounts of adsorbed gas exist, the ultra-low permeability of the shale matrix limits its effective extraction. CO_2 injection is one of the methods that are largely implemented for EOR purposes due to the availability of CO_2, the economics of operation, specific properties of CO_2 gas, and positive environmental impact. CO_2 can be used for enhanced gas recovery as well [60]. The process of EGR (enhanced gas recovery) using CO_2 is mainly dominated by pressurizing effect. The pressurizing effects can cause CO_2 injection to increase the rock permeability. The amount of CO_2 injected into the well will be divided into two amounts; about 1% of injected CO_2 will be produced, while 99% of injected CO_2 will be stored in the reservoir. Therefore, tight shale gas reservoirs potentially make excellent repositories for CO_2 sequestration purposes as well [60].

Various factors affect the recovery from tight shale reservoirs such as matrix porosity and permeability, hydraulic fracture half-length, and well spacing [61]. CO_2 injection in shales is often conducted using huff and puff method. The supercritical carbon dioxide injection repressurizes the reservoir after the initial production period. Once the injected gas soaks from the fractures into the shale's organic matrix through diffusion and convection, methane is released by the competitive adsorption since the shale has a stronger affinity for carbon dioxide than for methane. Then, during the second production period, the methane partial pressure is lowered and the shale gas production rate increases [62].

One example of such reservoirs is Chattanooga shale in Missouri, USA. The main objective of this project was to inject 500 tons of CO_2 to survey the injection and storage potential of CO_2 in a natural shale development while checking for enhanced gas recovery. The roads leading up to the wellpad were regraveled and graded to facilitate CO_2 delivery by truck to the injection site. The wellpad was cleared and graveled prior to moving equipment on site. A 70-ton CO_2 storage vessel was located permanently on site and refilled periodically by 20-ton tankers. The skid pump with all the controls and meters, as well as the propane tank and heater to heat the CO_2, was also located on site (**Figure 10**). The well head of the injection well was converted to accommodate the CO_2 injection by adding a gate valve, an inlet for the CO_2 line, as well as a tee for additional tests and monitoring [63].

The injection of 510 tons of CO_2 during this test exhibits the first successful injection of CO_2 in an organic shale formation to monitor for storage and enhanced gas recovery potential in Central Appalachia. This productive injection and monitoring of a CO_2 infusion in an organic shale reservoir are extraordinary

Figure 10.
Injection well site layout [63].

achievements and points of reference for CO_2-EGR and additionally geologic CO_2 storage in unconventional reservoirs. Once the well was brought back online after the soaking period, a significant increase in gas production occurred. During the first month of flowback, the average daily production rate was ~124 Mcf/day, which is over 8 times the average production for the last month before the well was taken offline for injection [64]. After 2 years of flowback, the well was still flowing at an increased production rate but is close to the projected historical production rate. The similar behavior of the injection well has been reported for modeling CO_2 'huff-and-puff' test in shale-oil reservoirs. The CO_2 concentration in the product gas has steadily declined during the flowback of the injection well and 41% of the injected CO_2 had been produced by the end of 2015 (17 months after flowback started). If the rate held constant, it would take over 8 years to produce all of the CO_2 injected [63, 64].

6.2 Tight oil reservoir

The tight oil reservoir is a type of unconventional oil reservoir that is hard to produce due to the low permeability. In recent years, the exploitation of tight oil reservoir increased due to advanced technologies in the production industry and high demand for the energy. Two main technologies that need to be increased are horizontal drilling and multi-stage hydraulic fracturing. U.S. Energy Information Administration reported that tight oil production will increase from 33% of total lower 48 onshore oil production to 51% in 2040 [59, 60]. However, the decline curves of primary production are steep due to low permeability [60, 66]. CO_2-EOR is utilized more commonly as compared to water flooding in case of tight oil reservoirs due to the poor sweep efficiency of water flooding and low injectivity of water in tight oil reservoirs. Moreover, in case of the reservoirs with a higher wetting tendency toward oil (oil-wet), water flooding would be less effective. CO_2-EOR could be implemented as continuous CO_2 injection or huff and puff technique. Hydraulic fracturing is one of the most important mechanisms in recoveries from tight oil reservoirs. The geometry, the number, and the spacing of the fracture can affect the recovery from the tight oil reservoirs. Bakken reservoir in the US is one of the largest unconventional tight oil reservoirs that has been produced since early 1950s. Through years of production and study of the reservoir, engineers decided that the following strategies are suitable for this field as shown in **Table 2**.

The followings outcomes were observed during the simulation of CO_2-EOR for the period of 30 years on Bakken Field:

1. Oil recovery factor increases with the increasing number of cycles of CO_2 huff and puff, and the incremental oil recovery factor at a 30-year period is 2.43% corresponding to three cycles in this case study.

2. Lower permeability, longer fracture half-length, and more heterogeneity are much favorable for the CO_2 huff and puff process.

3. The CO_2 diffusion mechanism is more pronounced than the convention mechanism for the reservoir with lower permeability during the CO_2 huff and puff process.

6.3 Heavy oil reservoir

Some of the world's largest reserves are heavy oil reservoirs. With oil in place equal to the largest conventional oil fields in the Middle East, these large reserves are found in more than thirty countries around the globe, but few of these deposits have been developed extensively. One of the problems in the heavy oil reservoir is asphaltenic oil precipitate in the reservoir. Asphaltene is a component in petroleum, especially heavy oils. The asphaltene content could be defined by its solubility. Basically, any component that dissolves in toluene and precipitates in alkane is considered as asphaltene [65]. Various strategies and possible action to oil recoveries are summarized in **Table 2**.

Overall, heavy oil reservoirs are considered not favorable for CO_2 flooding compared to light oil reservoir due to lack of sweep efficiency. The considerable viscosity difference between heavy oil and the injected CO_2 results in poor sweep efficiency from heavy oil reservoirs. Moreover, there is a possibility of asphaltene precipitation during miscible displacement; therefore, the compatibility of fluids is an important parameter to consider when designing EOR process for heavy oil reservoirs.

Strategies	Actions
Development of CO_2-EOR techniques for unlocking resources in tight oil formations	• CO_2 injection may be the most suitable technique for recovery. • Develop pragmatic and sustainable techniques in order to unlock resources in tight oil formations via this injection.
Tight formation characterization and multistage fracturing optimization	• For tight formations, the fracture systems are required such natural and hydraulic fracture. • The results can be used to optimize the well spacing and fracturing strategies.
Unlocking Bakken potential through CO_2 huff and puff	• Efficient in fractured reservoirs. • Potential to use mixtures such as CO_2/field gas, CO_2/methane, or CO_2/propane to improve the oil recovery under different conditions.
Optimum CO_2 and water-based oil recovery processes for Bakken formations	• To examine the oil recovery mechanisms and to evaluate the overall performance of WAG, GAW, and SWAG processes, in comparison to CO_2 flooding under miscible conditions. • Aim to determine the proper timing and also the slug sizes for different CO_2 and water-based oil recovery processes, including their effects on total oil recovery.

Table 2.
Strategies and action that could efficiently lead to increase in the recovery.

DOI: http://dx.doi.org/10.5772/intechopen.89540

7. Future horizons of CO_2-EOR and sequestration

The potential effect of CO_2-EOR is not so much a matter of whether but of when. The science of the process is known to the experts, and there are plenty of CO_2 emitters from different sources available and many oil reservoirs to host these CO_2 from these emitters. The key point is the economic decision that depends primarily on several factors. Oil price, capital cost of infrastructures, tax policies on CO_2 emission, cost of CO_2 capture from CO_2 emitters, and the speed of technology developments are the key parameters for economic decisions. These parameters could be hard to predict; however, developing a platform to facilitate the decision-making could speed up CO_2-EOR and sequestration significantly.

Author details

Erfan Mohammadian[1*], Badrul Mohamed Jan[2], Amin Azdarpour[3], Hossein Hamidi[4], Nur Hidayati Binti Othman[1], Aqilah Dollah[1], Siti Nurliyana Binti Che Mohamed Hussein[1] and Rozana Azrina Binti Sazali[1]

1 Faculty of Chemical Engineering, Universiti Teknologi Mara (UiTM), Shah Alam, Malaysia

2 Department of Chemical Engineering, University of Malaya, Kuala Lumpur, Malaysia

3 Department of Petroleum Engineering, Marvdasht Branch, Islamic Azad University, Marvdasht, Iran

4 School of Engineering, King's College, University of Aberdeen, Aberdeen, UK

*Address all correspondence to: erfan.723@gmail.com

References

[1] National Enrgy Technoogy Laboratory (NETL). Carbon Dioxide Enhanced Oil Recovery (Untapped Domestic Energy Supply and Long Term Carbon Storage Solution) [Internet]. 2010. Available from: http://www.netl.doe.gov/index.html

[2] Stalkup FI. Miscible Displacement, SPE Monograph No. 8. New York: Society of Petroleum Engineering of AIME; 1983

[3] Stalkup F. Miscible Displacement, Second Printing. Dallas, New York: Society of Petroleum Engineers of AIME; 1984

[4] Kokal S, Al-kaabi A. Enhanced Oil Recovery: Challenges & Opportunities. World Petroleum Council: Official Publication; 2010

[5] Stosur GJ. EOR: Past, present and what the next 25 years may bring. In: SPE Paper 84864, Presented at the SPE IOR Conference in Asia Pacific; October 20-21, 2003; Kuala Lumpur, Malaysia 2003

[6] Thomas S. EOR—An overview. Oil and Gas Science and Technology. Revue de l"IFP. 2008;**63**(1):9-19

[7] IPCC AR5 WG2 A. Field CB et al., editors. Climate Change 2014: Impacts, Adaptation, and Vulnerability. Part A: Global and Sectoral Aspects. Contribution of Working Group II (WG2) to the Fifth Assessment Report (AR5) of the Intergovernmental Panel on Climate Change (IPCC). Cambridge University Press, Archived from the original on 28 April 2016. Archived; 2014

[8] Saxon J Jr, Breston JN, MacFarlane RM. Laboratory tests with carbon dioxide and carbonated water as flooding mediums. Production Monthly. 1951;**16**:8

[9] Beeson DM, Ortloff GD. Laboratory investigation of the water-driven carbon dioxide process for oil recovery. Transactions of AIME. 1959;**216**:388-391

[10] Holm LW, Josendal VA. Effect of oil composition on miscible-type displacement by carbon dioxide. Society of Petroleum Engineers Journal. 1982;**22**(2):87-98

[11] Kane A. Performance review of a large scale C0,-WAG enhanced recovery project. SACROC Unit-Kelly Snyder Field. Journal of Petroleum Technology. 1979;**31**(2):217-231

[12] Mathiassen OM. CO_2 as Injection Gas for Enhanced Oil Recovery and Estimation of the Potential on the Norwegian Continental Shelf. Stavanger: NTNU—Norwegian University of Science and Technology; 2003

[13] Teklu TW, Alharthy N, Kazemi H, Yin X, Graves RM. Hydrocarbon and non-hydrocarbon gas miscibility with light oil in shale reservoirs. Society of Petroleum Engineers. 2014. DOI: 10.2118/169123-MS

[14] Morishige K, Fujii H, Uga M, Kinukawa D. Capillary critical point of argon, nitrogen, oxygen, ethylene, and carbon dioxide in MCM-41. Langmuir. 1997;**13**(13):3494-3498

[15] Travallonia L, Castierb M, Tavaresa FW, Sandler SI. Critical behavior of pure confined fluids from an extension of the Van der Waals equation of state. Journal of Supercritical Fluids. 2010;**55**(2):455-461

[16] Zarragoicoechea GJ, Kuz VA. Critical shift of a confined fluid in nanopore. Fluid Phase Equilibria. 2004;**221**(2004):7-9

[17] Ahmadi K, Johns RT. Multiple mixing cell method for

MMP calculation. SPE Journal. 2011;**16**(4):733-742

[18] Graue DJ, Zana ET. Study of a possible CO_2 flood in the Rangely Field, Colorado. Journal of Petroleum Technology. 1981;**1981**:1312

[19] Verma BMK, Jewell S, Survey USG. Fundamentals of Carbon Dioxide-Enhanced Oil Recovery (CO2-EOR)—A Supporting Document of the Assessment Methodology for Hydrocarbon Recovery Using CO2-EOR Associated with Carbon Sequestration. 2015

[20] Friedmann SJ. Carbon Capture and Sequestration Technologies: Status and Future Deployment. UCRL-BOOK-235276. 2007

[21] Kang P, Lim J, Huh C. Screening criteria and considerations of offshore enhanced oil recovery. Energies. 2016;**9**(1):44. DOI: 10.3390/en9010044

[22] Bachu S. Sequestration of CO_2 in geological media in response to climate change: Roadmap for site selection using the transform of the geological space into the CO_2-phase space. Energy Conversion and Management. 2002;**43**:87-102

[23] Webster CM, Zhang Y, Park D, Gibbons K. Workflows & Considerations for CO_2 Injection in a Highly Depleted Gas Field in the Southern North Sea. In Offshore Europe. Society of Petroleum Engineers; 2011

[24] Kalantari-dehghani A, Gholami V, Moghadasi J, Abdi R. Formation Damage Through Asphaltene Precipitation Resulting From CO_2 Gas Injection in Iranian Carbonate Reservoirs. SPE SPE-99631-PA. 2008. ISBN 1930-1855

[25] Chadwick RA, Williams GA, Williams JDO, Noy DJ. Measuring pressure performance of a large saline aquifer during industrial-scale CO injection: The Utsira sand. International Journal of Greenhouse Gas Control. 2012;**10**:374-388. ISSN 1750-5836

[26] Zagreb. Design and mechanical integrity of CO_2 injection. Wells. 2011;**23**:1-8

[27] Wei N, Li X, Dahowski RT, Davidson CL, Liu S, Zha Y. Economic evaluation on CO 2 -EOR of onshore oil fields in China, International Journal of Greenhouse Gas Control. 2015;**37**:170-181. DOI: 10.1016/j.ijggc.2015.01.014

[28] Fukai I, Mishra S, Moody MA. Economic analysis of CO_2-enhanced oil recovery in Ohio : Implications for carbon capture, utilization, and storage in the Appalachian Basin region. International Journal of Greenhouse Gas Control. 2016;**52**:357-377. DOI: 10.1016/j.ijggc.2016.07.015

[29] Verma BMK, Jewell S, Survey USG. Fundamentals of Carbon Dioxide-Enhanced Oil Recovery (CO_2-EOR)—A Supporting Document of the Assessment Methodology for Hydrocarbon Recovery Using CO_2-EOR Associated with Carbon Sequestration

[30] Webster CM, Zhang Y, Park D, Gibbons K. Workflows & Considerations for CO_2 Injection in a Highly Depleted Gas Field in the Southern North Sea. Offshore Europe: Society of Petroleum Engineers; 2011

[31] Chadwick RA, Williams GA, JDO W, Noy DJ. Measuring pressure performance of a large saline aquifer during industrial-scale CO injection: The Utsira Sand, Norwegian North Sea. International Journal of Greenhouse Gas Control. 2012;**10**:374-388. ISSN: 1750-5836

[32] Macon BR. Design and Operation of the Level Land Unit CO_2 Injection Facility SPE Annual

Technical Conference and Exhibition. SPE-8410-MS

[33] Saadawi H. Design of Surface Facilities for a High-Pressure Miscible Gas Injection Pilot Scheme Macintyre. 1989

[34] Alderman D, McFarland R, Mcintyre J, Minissale J. Field facilities for CO_2 tertiary floods-a conceptual design. In: 57th Annual Fall Technical Conference and Exhibition of the Society of Petroloum Engineers of AIME. 1982. pp. 1-7

[35] Iwasaki S, Kamijo T, Takashina T, Tanaka H. Large scale flue gas CO_2 recovering/CO_2 cost and total system for CO_2 enhanced oil recovery. Technical Review. 2004;**41**(4):1-6

[36] Linn. CO_2 Injection and Production Field Facilities Design Evaluation and Considerations. 1987

[37] Macintyre KJ. Design considerations for carbon dioxide injection facilities. Petroleum Society of Canada. 1986. DOI: 10.2118/86-02-09

[38] Research report: Hazard Analysis for Offshore Carbon Capture Platforms and Offshore Pipelines. London: Energy Institute and Global CCS Institute; 2017

[39] Donghyun YK, Yi HS, Dong-Hun K, Jin-Kuk K. Feasibility study and process development of CO_2 EOR surface facilities. 2015

[40] IEAGHG. Effects of Impurities on Geological Storage of CO_2. 2011

[41] Kikuta K, Hongom S, Tanase D, Ohsumi T. Field test of CO_2 injection in Nagaoka, Japan. In: 7th International Conference on Greenhouse Gas Control Technologies. 2004

[42] Goodyear SG, Koster MP, Marriott KA, Paterson A, Sipkema AW, Young IM, Production BV. Moving CO 2 EOR Offshore. 2011

[43] Verma MK. 2015 Fundamentals of carbon dioxide-enhanced oil recovery (CO_2-EOR)—A supporting document of the assessment methodology for hydrocarbon recovery using CO_2-EOR associated with carbon sequestration: U.S. Geological Survey Open-File Report 2015-1071. 19 p. DOI: 10.3133/ofr20151071

[44] ISSN 2331-1258 (online) IPCC (2014). International panel on Climate Change. Berlin, Germany: Intergovernmental Panel on Climate Change. 11th Assessment Report; IPCC Secretaria; 2014

[45] Allen MR et al. Framing and context. In: Masson-Delmotte VP. Zhai H-O, Portner D, Roberts J, Skea PR, Shukla A, et al, editors. Global Warming of 1.5°C. An IPCC Special Report on the Impacts of global Warming of 1.5°C above Pre-Industrial Levels and Related Global Greenhouse Gas Emission Pathways, in the Context of Strengthening the Global Response to the Threat of Climate Change, Sustainable Development, and Efforts to Eradicate Poverty. Geneva, Switzerland: World Meteorological Organization; 2018. pp.49-91

[46] Da Silva PR, Ranjith PG. A study of methodologies for CO_2 storage capacity estimation of saline aquifers. Journal of Fuel. 2011;**93**:13-27

[47] Bradshaw J, Bachu S, Bonijoly D, Burruss R, Holloway S, Christensen N. CO_2 storage capacity estimation: Issues and development of standards. International Journal of Greenhouse Gas Control. 2007;**1**(1):62-68

[48] IEA Energy Technology Essentials. IEA Annual Report. Paris, France: 2012

[49] Hassanzadeh H. Mathematical Modeling of Convective Mixing in Porous Media for Geological CO_2 Storage [PhD thesis]. Canada: University of Calgary; 2006

[50] Zou CN, Zhu RK, Liu KY. Tight gas sandstone reservoirs in China: Characteristics and recognition criteria. Journal of Petroleum Science and Engineering. 2012;**88-89**:82-91

[51] Gozalpour F, Ren SR, Tohidi B. 2005 CO_2 EOR and storage in oil reservoirs. Oil and Gas Science and Technology. Revue de l"IFP. 2005;**60**(3):537-546

[52] Rogala A, Ksiezniak K, Jan K, Jan H. Carbon dioxide sequestration during shale gas recovery. Physicochemical Problems of Mineral Processing. 2014;**50**(2):681-692

[53] Senger K, Tveranger J, Alvar B, Olaussen S, Ogata K, Larsen L. CO_2 storage resource estimates in unconventional reservoirs: Insights from a pilot-sized storage site in Svalbard, Arctic Norway. Environment and Earth Science. 2014;**73**(8):3987-4009. DOI: 10.1007/s12665-014-3684-9

[54] Cudjoe S, Vinassa M, Henrique Bessa Gomes J, Barati RG. A comprehensive approach to sweet-spot mapping for hydraulic fracturing and CO_2 huff-n-puff injection in Chattanooga shale formation. Journal of Natural Gas Science and Engineering. 2016;**33**:1201-1218. DOI: 10.1016/j.jngse.2016.03.042

[55] Louk K, Ripepi N, Luxbacher K, Gilliland E, Tang X, Keles C, et al. Monitoring CO_2 storage and enhanced gas recovery in unconventional shale reservoirs: Results from the Morgan County, Tennessee injection test. Journal of Natural Gas Science and Engineering. 2017;**12**:2355-2388. DOI: 10.1016/j.jngse.2017.03.025

[56] Jong EC, Restrepo JA, Luxbacher KD, Kirsch PA, Mitra R, Hebblewhite BK, et al. Risk management: Adapting riskgate for underground coal mines in the United States. Mining Engineering. 2016;**68**(3):51-57. DOI: 10.19150/me.6502

[57] U.S. Energy Administration. Monthly crude oil and natural gas production. Monthly Report. 2017.

[58] Zuloaga-Molero P, Yu W, Xu Y, Sepehrnoori K, Li B. Simulation study of CO_2-EOR in tight oil reservoirs with complex fracture geometries. Scientific Reports. 2016;6: Article number: 33445

[59] Pilcher RS, McDonough Ciosek J, McArthur K, Hohman JC, Schmitz P. Ranking production potential based on key geological drivers—Bakken case study. International Petroleum Technology Conference. 2011. DOI: 10.2523/IPTC-14733-MS

[60] Dechongkit P, Prasad M. Recovery factor and reserves estimation in the Bakken petroleum system (analysis of the antelope, Sanish and Parshall fields). Society of Petroleum Engineers. 2011. DOI: 10.2118/149471-MS

[61] Kumar S, Hoffman T, Prasad M. Upper and lower Bakken shale production contribution to the middle Bakken reservoir. In: Unconventional Resources Technology Conference. 2013

[62] Huang T, Yang H, Liao G, Zeng F. Optimization of CO_2 flooding strategy to enhance heavy oil recovery. In: Paper Presented at the Society of Petroleum Engineers—SPE Canada Heavy Oil Technical Conference 2015, CHOC 2015. 2015. pp. 980-999

[63] Rojas GA, Zhu T, Dyer SB, Thomas S, Ali SF. Scaled model studies of CO_2 floods. Society of Petroleum Engineers Reservoir Engineering.

1991;**6**(2):169-178. DOI: 10.2118/18083-PA

[64] Chukwudeme EA, Hamouda AA. Enhanced Oil Recovery (EOR) by Miscible CO_2 and Water Flooding of Asphaltenic and Non-Asphaltenic Oils. 2009

[65] Liao Z et al. Discussion on the structural features of asphaltene molecules. Energy & Fuels. 2009;**23**(12):6272-6274

[66] National Energy Technology Laboratory. Carbon Dioxide Enhanced Oil Recovery Report. USA: U.S. Department of Energy; 2010